Messerschmitt Me 262:
Variations, Proposed Versions
& Project Designs Series

Messerschmitt Me 262:
Variations, Proposed Versions
& Project Designs Series

Design Concept, Prototypes, V Series, Flight Tests

David Myhra

Schiffer Military History
Atglen, PA

Book Design by Ian Robertson.

Printed in China.
ISBN: 0-7643-1888-8

We are interested in hearing from authors with book ideas on military topics.

Published by Schiffer Publishing Ltd.
4880 Lower Valley Road
Atglen, PA 19310
Phone: (610) 593-1777
FAX: (610) 593-2002
E-mail: Info@schifferbooks.com.
Visit our web site at: www.schifferbooks.com
Please write for a free catalog.
This book may be purchased from the publisher.
Please include $3.95 postage.
Try your bookstore first.

In Europe, Schiffer books are distributed by:
Bushwood Books
6 Marksbury Avenue
Kew Gardens
Surrey TW9 4JF
England
Phone: 44 (0) 20 8392-8585
FAX: 44 (0) 20 8392-9876
E-mail: Bushwd@aol.com.
Free postage in the UK. Europe: air mail at cost.
Try your bookstore first.

Contents

Acknowledgments

Clockwise from above left:
Betsy Hertel
Marek Rys
Günter Sengfelder
Mario Merino
Emil Petrinic
Jozef Gatial

Introduction

Numerous books have been written over the years about the truly revolutionary twin Junkers Jumo 004 gas turbine powered Messerschmitt Me 262A-1a pursuit fighter. This flying machine was most significant because it was the first gas turbine powered airplane to go from the drawing board, prototype testing, series production, and then on to combat duty. Sure, England, America, and Japan had gas turbine powered flying machines around the same time, but no other country had advanced them to series production, much less combat status.

Willy Messerschmitt, on his 80th birthday, was asked which of his many aircraft was he most proud of creating during his long aviation career. He answered the Me 262. He went on to explain the obvious: "work on this design provided the aviation world with such a rich wellspring of knowledge...a role model for aircraft designs appearing in the Free World and Communist countries." Other aircraft designers throughout the world would agree that the ultimate configuration of the Me 262 was considered "radical thinking" even though only 1,433 Me 262 were assembled by war's end another 500 were in various stages of assembly stymied due to the lack of components. All in all, probably only 600 to 800 Me 262s ever reached Luftwaffe fighting units.

The Me 262 was revolutionary, even though the number of units produced is extremely small, in another significant way, too. When it entered combat as a pursuit fighter in July 1944, it forced all existing piston powered flying machines of World War Two into immediate obsolescence. This was because the Me 262 was a full 125 miles per hour [200 kilometers/hour] faster than any conventional powered aircraft. The maximum speed of Germany's Focke-Wulf Fw 190D, their fastest piston powered fighter, was 426 miles per hour [686 kilometers/hour] at 21,653 feet altitude where it performed best. But the forward speed of the standard Me 262 A-1a was 540 miles per hour [869 kilometers/hour] at 19,684 feet altitude where it, too, performed best. Plus, the Me 262 could surpass the speed of sound in a dive. With its fledgling and temperamental gas turbines the Me 262 showed the world that, with anticipated advances in gas turbine technology, speeds well in excess of Mach One would be occurring on a regular basis by the late 1940s.

Me 262 A-1a. Scale model and photograph by Günter Sengfelder.

Me 262 A-1a. Scale model and photograph by Günter Sengfelder.

Given the Me 262's superior speed, aircraft design planners at the RLM and Messerschmitt AG began considering the various ways the Me 262 could be put use in addition to a pursuit fighter. Planners from the RLM and Messerschmitt AG were coming up with numerous follow-on developments for the Me 262. Follow-on developments is what this series is all about.

It is believed that by war's end in Europe on May 8th 1945, real and planned Me 262 variations/versions may have reached 100 or more. These variations/versions included designs equipped with multiple machine cannoned fighters, R4M rockets, Wr.Gr 210 aerial mortars, aerial torpedoes, and Henschel Hs 293 guided missile carriers. There were Me 262 variations/versions for all weather fighters, close-in ground support flying machines, short and long range tactical aerial reconnaissance flying machines...both armed and unarmed. There were plans for heavy bomber interceptors and ultra high speed bombing machines powered by bi-fuel liquid rocket engines. There was a design which include two Eugen Sänger designed Lórin ramjets each mounted above the Me 262's wing. Other Me 262 variations/versions planned included radar equipped night fighters to be powered by the advanced high thrust Heinkel-Hirth HeS 011 axial flow gas turbine and the turboprop version known as the DB-021. The HeS 011 gas turbine was proposed to power high speed, high altitude Me 262 fighters with up to 45 degree swept back wings. The HeS 011 gas turbines would be tucked in internally at the wing root. Several other Me 262 proposed designs called for the cockpit to be placed just aft the nose cone. Other designs had the Me 262's cockpit located far aft actually at the base of the vertical stabilizer.

Featured in this first book are early Me 262 design concepts and early wind tunnel test models. Also included are all the Me 262 "Versuchs" research or pre-production flight testing machines as well as pre-production "Serien" or S-type flight test machines. The "S" types were added to the initial flight test "Versuchs" because Messerschmitt AG factory test pilots were destroying them on a regular basis...so the RLM required more. Me 262 design concepts, Versuchs, and S-types. Later books in the series will include the Me 262 A-1a standard production fighter and all the known Me 262 "A/U" and "B/U" fighter, bomber, and night fighter versions. The next volume will include all the known Me 262 C series of proposed high speed, high altitude bomber interceptors. The Me 262 C series would have included those Me 262s with auxiliary bi-fuel liquid rocket motors to augment its climbing ability to reach and hunt down the high flying Boeing B-17 bombers and the RLM's belief that the Boeing B-29 was soon to appear over Germany, too. Also featured with the C series are the Me 262 HG (home defender) series of high speed, high altitude fighter designs to be powered by the advanced Junkers Jumo 004D and Heinkel-Hirth HeS 011 gas turbines. Several of these designs included 35 and 45 degree wing sweep. The final volume will include all of the Me 262 wrecks found post war. Although 1,400 Me 262s were assembled by the time the hostilities ended on May 8th 1945, only a handful of Me 262s remained flight ready, and they were mostly taken by the victorious American, British, French, and Soviet armies. However, most of the Me 262 were destroyed by Allied bombing, cannibalized for parts, or found abandoned throughout Germany and surrounding countries. Hundreds more were found in various stages of assembly in forests and caves. Another volume will feature nearly three hundred photos of Me 262 wrecks found post war.

First impression of the Me 262 by USAAF Intelligence from 1943 and based on visual accounts.

An updated impression of the Me 262 by USAAF Intelligence from December 1944. This image is very accurate because it is based on gun camera images from North American P-51 and Republic P-47 fighters.

Me 262 A-1a. Scale model and photograph by
Günter Sengfelder.

Me 262 A-1a. Scale model and photograph by
Günter Sengfelder.

The United Kingdom's first operational twin (centrifugal) gas turbine powered fighter the Gloster Meteor. This F-1 flying machine was used sparingly inside England, and only then to chase down the Fi 103 V1 "buzz bombs" launched against England from bases in France.

America's Bell P-59A "Aircomet" twin centrifugal gas turbine powered prototype. Its first flight test occurred on September 31st 1942, and the first production P-59As were delivered to the USAAF in August 1944.

Preliminary Design Ideas Leading to the Me 262 A-1a

In late 1938, the Reichsluftministerium (RLM) or German Air Ministry issued a contract to Messerschmitt AG-Augsburg for a flying machine to be powered by two BMW P-3302 axial flow gas turbine engines each rated at 1,322 pounds [600 kilograms] thrust. The RLM designated the flying machine as Project 1065.

The reasons why the RLM gave the contract to Messerschmitt and not Heinkel AG, or both, is unclear. Perhaps it was due to Willy Messerschmitt's world-wide reputation for designing outstanding record breaking flying machines, such as his world air record speed holder Me 209 V1.

Dr.-Ing. Hermann Oestrich of BMW had assured the RLM that their Project 3302 gas turbine would be ready for field testing at the end of 1939, in an air frame of RLM's choice.

By June 7 th 1939, Messerschmitt AG had completed three preliminary designs for a flying machine powered by twin gas turbine engines. Messerschmitt AG was referring to the three as Project Design 65. They submitted all three designs in a single package to the RLM for their evaluation.

Messerschmitt AG individuals directly involved in the design of the Project 65 included:

- Willy Messerschmitt - owner of Messerschmitt AG;
- Woldemar Voigt - head of the Projects Office;
- Wolfgang Degel - deputy head of the Projects Office;
- Karl Althoff - designer in the Projects Office;
- Walter Eisenmann - deputy head of the Aerodynamics Department;
- Riclef Schomerus - head of the Aerodynamics Department;

When Willy Messerschmitt and his hand picked Project 65 design started their thinking about an airframe design to be powered by twin axial flow gas

Willy Messerschmitt. Taken in the early 1940s.

Dipl.-Ing. Woldemar Voigt. Design genius of Messerschmitt AG and largely responsible for the design of the Me 262. Taken in late 1939.

Willy Messerschmitt (left) and Wolfgang Degel (right), deputy head of the Projects Office under Woldemar Voigt.

von Chlingensperg (left) and Riclef Schomerus (right) head of Mersserschmitt AG's Aerodynamics Department. Both men lost their lives on February 9th 1945, on their way to Japan in U864 when it was torpedoed and sunk by the British submarine "Venturer" off the coast of Bergen. Both men were to deliver drawings/components of the Me 262A-1a to the Japanese to help them build their version, known as the Nakajima "Kikka."

turbines, several general planforms were already established. Messerschmitt himself had expressed his vision for this revolutionary new airframe in what is being called his hand drawn sketch (side port profile) from October 17th 1939. Beyond the hand drawn sketch, Messerschmitt's team would go through as many as twelve (perhaps even more unknown planforms) before settling on what would become the Messerschmitt AG Project Design 262 V1. The evolution of designs beginning with Willy Messerschmitt's hand drawn design includes the following and they are presented in greater detail with digital images in the following pages:

- Willy Messerschmitt's hand drawn sketch of October 17th 1939;
- Messerschmitt Project Design 65 featuring a straight low wing with oval fuselage and BMW P-3302 gas turbines through the wing;
- Messerschmitt Project Design 65 featuring a straight mid wing with oval fuselage and BMW P-3302 gas turbines through the wing;
- Messerschmitt Project Design 65 wind tunnel drawing from July 14th 1940;
- Messerschmitt Project Design 65 featuring a straight low wing triangular fuselage and BMW P-3302 gas turbines through the wing;
- Messerschmitt Project Design 262 Modell I featuring a low wing triangular fuselage and BMW P-3302 gas turbines below the wing;
- Messerschmitt Project Design 262 evolving Post Modell I featuring a full 35 degree sweep low wing AVA wind tunnel model design;
- Messerschmitt Project Design 262 featuring a nose wheel;
- Messerschmitt Project Design 1070 featuring a swept back low wing oval fuselage nose wheel, BMW P-3302 gas turbines through the wing;
- Messerschmitt Project Design 262 VI - Stage One
- Messerschmitt Project Design 262 VI - Stage Two
- Messerschmitt Project Design 262 VI - Stage Three

There is some discrepancy surrounding Project 65. It appears that Messerschmitt AG referred to their BMW P-3302 powered airframe as the P-65. It appears that the RLM preferred to identify the design as Project 1065. Not much of a difference, really. Nonetheless, the officials of Messerschmitt AG, especially Willy Messerschmitt himself, referred to the design as Project 65...perhaps just a form of abbreviation. At any rate, there is no doubt that the designation Project 65 or Project 1065 in the documents of the time refer to the same thing.

A rare photo of Luftwaffe chief Hermann Göring smiling as he watches a demonstration flight of the Me 262 V6 at Lager Lechfeld. The Me 262 V6 was powered by twin Junkers Jumo 004A gas turbines. To Göring's right is Willy Messerschmitt. At the far right is Dr.-Ing. Anselm Franz of Junkers Jumo...the man who perfected the Jumo 004 gas turbine after taking over research and development started by the engineering genius Dr.-Ing Professor Herbert Wagner in the late 1930s.

About mid-June 1939, the RLM awarded Messerschmit AG a contract to construct a mockup of their proposed Project 1065. Then, on December 19th 1939, the RLM inspected a mock up of Project 1065 prepared by Messerschmitt AG. On March 1st 1940 the RLM awarded Messerschmitt AG a contract to construct three Project 1065 prototypes. Several details were still unclear. For example:

- Messerschmitt AG's engineers' original design submitted to the RLM called for placing the BMW P-3302 gas turbines inside the wings through an "eye" in the wing's main spar.
- BMW's Hermann Oestrich announces that the overall diameter of their P-3302 gas turbine has increased beyond original design plans.

- Messerschmitt AG's engineers conclude that the greater diameter of the BMW P-3302 gas turbine will no longer fit inside the wing on their P-1065, but will now have to be hung beneath the wing.
- The design to hang the BMW P-3302 beneath the wing requires considerable redesign of P-1065. BMW must also strengthen certain parts of their gas turbine so it can be hung from the P-1065's main wing spar. Heavier attachment points, a stronger wing, plus a redesigned fuselage, and so on now mean that the P-1065 is a much heavier flying machine than originally anticipated.
- Messerschmitt AG submits to the RLM their modified design to carry both BMW P-3302 gas turbines beneath the wing sometime between April 15th 1940 and May 15th 1940. Sources disagree.
- In early July 1940, the RLM accepts Messerschmitt AG's redesign of Project 1065.

Willy Messerschmitt's Hand-Drawn Sketch of October 17th 1939 for a Gas Turbine-Powered Pursuit Fighter and a Tricycle Landing Gear

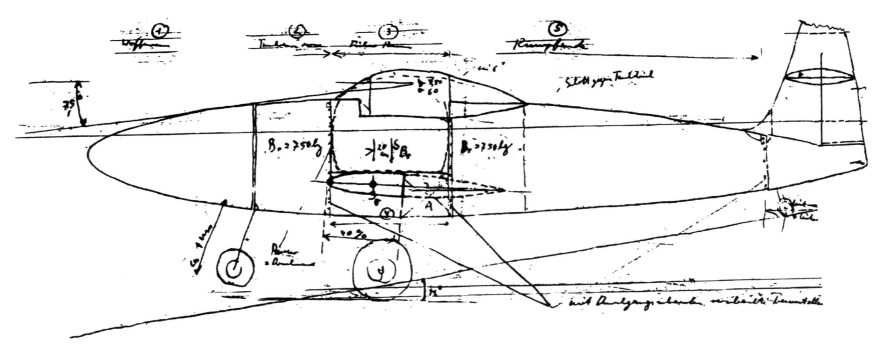

A free-hand sketch of his thoughts for a gas turbine powered fighter from October 17th 1939, by Willy Messerschmitt. This is earliest known drawing of what would later become the Me 262 A-1a. Messerschmitt's sketch features: oval shaped fuselage, mid wing location, tricycle landing gear, and the horizontal tailplane located half way up the vertical stabilizer.

Messerschmitt Project Design 65
Straight Low Wing with Oval Fuselage. BMW P-3302 Gas Turbines through the Wing

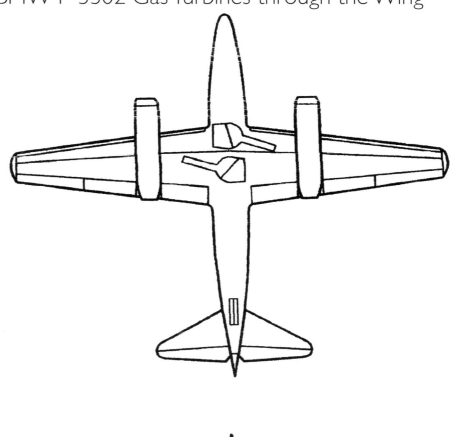

Woldemar Voigt's Project 65 team's first idea/configuration for a twin BMW P-3302 gas turbine powered fighter from late 1939. This tail wheel design featured a straight low wing with the gas turbines positioned within the wing. Main undercarriage had a staggered arrangement when it was retracted and was decided to be unacceptable due to gear attachment points and gear location and gear retraction. Voigt's team liked the low wing configuration but not the staggered gear. Idea abandoned due to complicated landing gear arrangement.

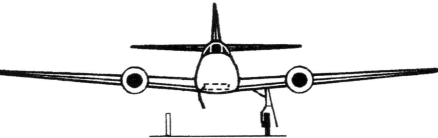

Messerschmitt Project Design 65
Straight Mid Wing with Oval Fuselage. BMW P-3302 Gas Turbines through the Wing

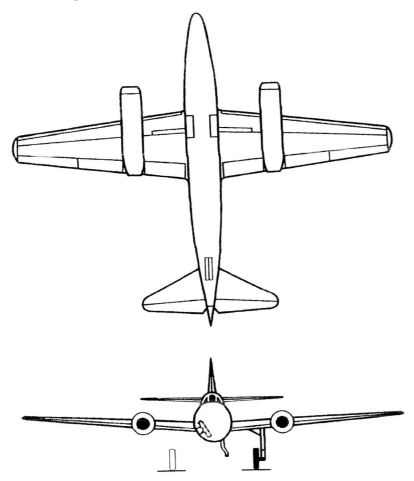

Dr.-Ing. Hermann Oestrich (1903-1973) struggled to perfect the BMW 002 and the BMW 003 axial flow gas turbines, suffering ridicule and embarrassment along the way, but kept at it until Germany's surrender on May 8th 1945.

Woldemar Voigts' Project 65 team's second idea/configuration for a twin BMW P-3302 gas turbine pursuit fighter. This proposed tail wheel layout features a straight mid-wing with the twin gas turbines located inside the wing. Voigt's team later determined that this design was unacceptable due to the fact that a long-legged landing gear would be required and the gear would have been stowed in the fuselage sides. Voigt's team was moving away from the mid-wing configuration plus they wanted the gear to retract flat inside the fuselage bottom. Idea abandoned due to the long landing gear struts and complications of stowing the gear.

Opposite:
Above: Hermann Oestrich's BMW P-3302 gas turbine as seen from its port side. Had it lived up to its promise this gas turbine not only would have powered the flying machines from Messerschmitt AG but the all-wing Horten Ho 9, too. Below: Pen and ink illustration of the BMW P-3302 gas turbine as seen from its port side.

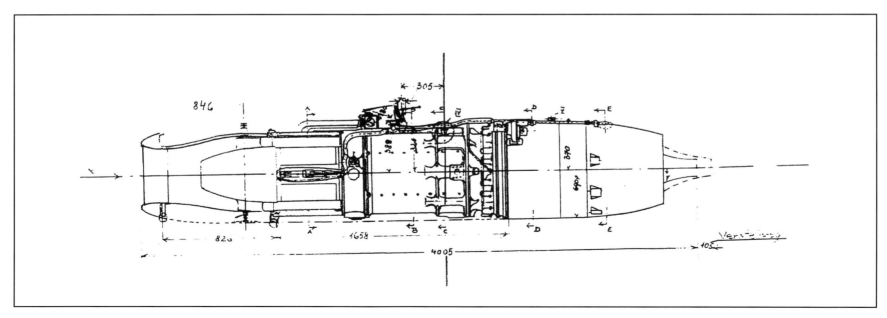

The BMW P-3302 gas turbine as seen from its nose port side.

The BMW P-3302 gas turbine as seen from its rear port side.

Voigt's Me P-65 design team threw out the mid-wing concept in favor of the low wing. In this design study from June 7th 1939, Voigt's group changed the main landing gear. Notice that main wheels and their retraction path are similar to the Me 262 A-1. This latest design also featured a wing identical to a Bf 109E's wing and was probably designed to carry twin BMW P-3304 gas turbines mounted through the wing. Courtesy: Me 262, Volume One, J. Richard Smith & Eddie J. Creek, Classic Publications, East Sussex, England, 1997.

Messerschmitt Project Design 65
Wind Tunnel Drawing without Gas Turbines by AVA, July 14th 1940

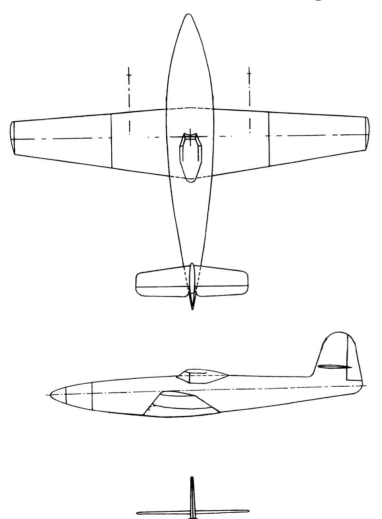

The Voigt team learned that BMW's P-3302 small diameter gas turbine was failing to meet its design thrust and would have to be re- engineered with a larger diameter compressor. The new BMW P-3304 gas turbines, though larger in diameter, might still fit inside the wing. Seeking alternatives on engine location if BMW was wrong, Voigt's Me P-65 team considered placing the gas turbines under the wing or on top of the wing. To determine the most suitable location for the gas turbines, Voigt took a scale model to the wind tunnels at Aerodynamische Veruschanstalt Aerodynamic (AVA) at Göttingen in 1940. AVA's aerodynamicists would start with a clean wing with no gas turbines.

Messerschmitt Project Design 65
AVA Wind Tunnel Model with BMW P-3302 Gas Turbines
Mounted above the Wing's Surface

One idea of the Voigt team was a Me P-65 with its BMW P- 3302/ 3304 axial flow gas turbines mounted above the wing as featured in this illustration believed to be from July 1940. Courtesy: Jet Planes of the Third Reich, Manfred Griehl, Monogram Aviation Publications, Sturbridge, MA, 1998.

A wind tunnel model of the Me P-65 being tested with proposed BMW P-3304 gas turbines mounted above the wings at AVA-Göttingen. About July 1940. This version is seen with the straight Bf 109E style wings and the new triangular-shaped fuselage.

Messerschmitt Project Design 65
AVA Wind Tunnel Model with BMW P-3302 Gas Turbines
Mounted High above the Wing's Surface

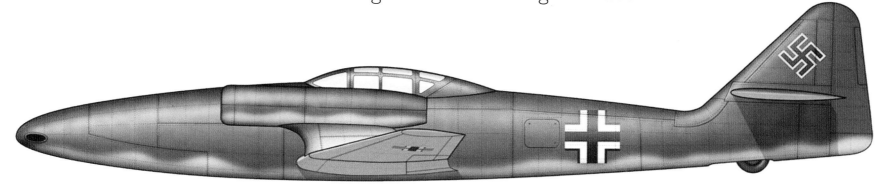

Digital image of the Me P-65 with its above wing gas turbines by Joself Gatial.

AVA-Göttingen also wind tunnel tested the Me P-65 with its gas turbines raised high above the surface of the wing. This photo features the gas turbine well above the starboard wing.

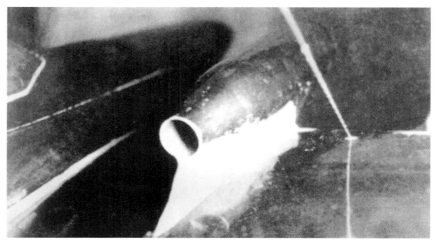

A starboard size view of the Me P-65 at the AVA-Göttingen wind tunnel with its gas turbine mounted in an above wing location. To the left of the photograph can be seen the Me P-65's cockpit windscreen.

Messerschmitt Project Design 65
Low Wing Straight Inner Wing with 18° Swept Outer Wing
Triangular Fuselage BMW P-3302 Gas Turbines through the Wing Spar

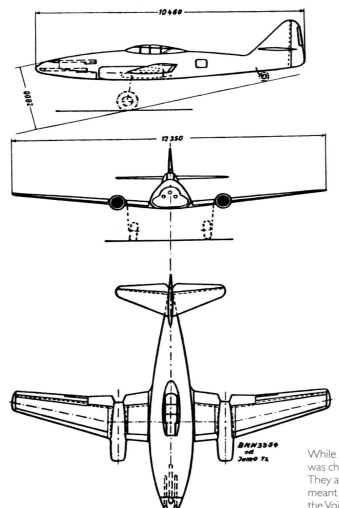

While AVA-Göttingen was wind tunnel testing the most efficient location for the Me P-65's gas turbines, Voigt's team was changing the shape of the project's wing. The triangular-shaped fuselage was the shape Voigt's team had settled on. They also needed to change the Project's center of gravity due to the heavier gas turbines coming from BMW, and this meant giving the wing a sweep back. This 3-view drawing, believed to be from March 1940, shows the initial thinking of the Voigt design team and their modest 18 degree wing sweep for its outboard wing. The gas turbine locations were still through the wing...which was no longer feasible due to their increasing diameter. The Me P-65 was still a tail dragger.

A direct overhead view of AVA-Göttingen's Me P-65 with its gas turbines mounted through the wing, however, featuring an 18 degree wing sweep on its outer wing.

The Me P-65's shape was coming closer to perfection, however, as this photo from the AVA-Göttingen wind tunnel shows, the placement of the twin gas turbines was still unsettled.

Messerschmitt Project Design 262 Modell I
Straight Low Wing Triangular Fuselage BMW P-3302 Gas Turbines below the Wing

The AVA-Göttingen Me 262 Modell I wind tunnel model featuring the twin gas turbines under the wing as well as the Bf 109E style straight wing.

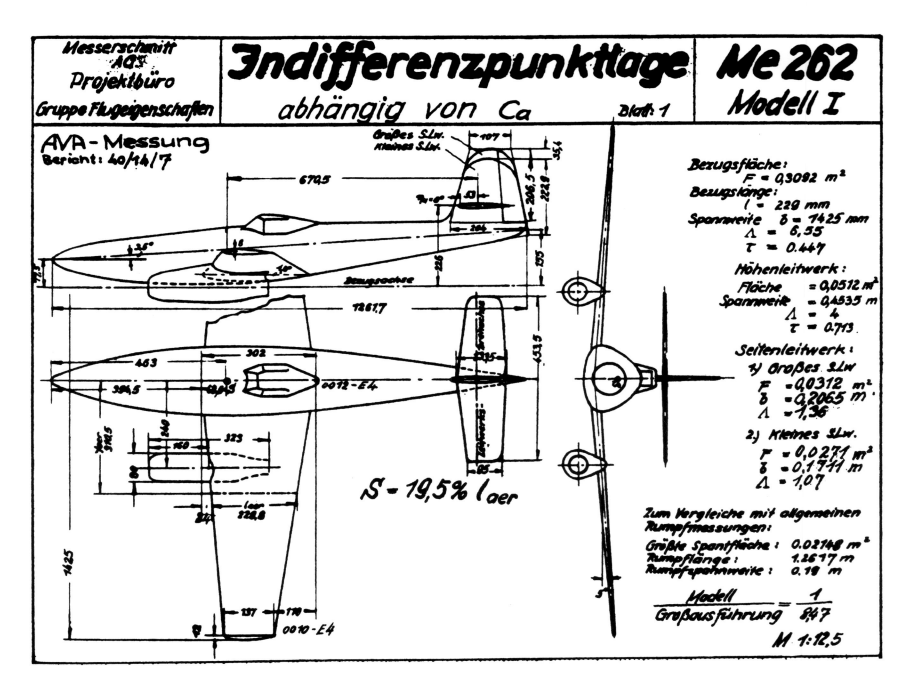

A clean three view of Messerschmitt AG's Me P-262 Modell I seen in the previous drawing.

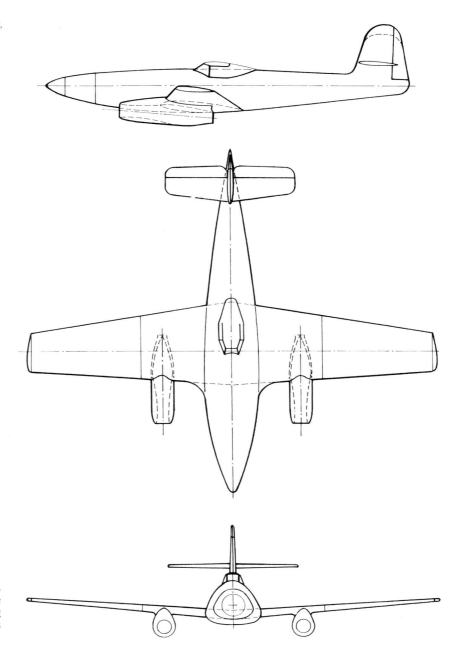

Messerschmitt P-65 is given a new designation. Notice that this drawing dated July 14th 1940, that Me P-65 is no longer being called Me P-65 but has been given a new designation...Me 262! Also noticeable from the drawing is the indication that its gas turbines are now hanging beneath the wing. Although this drawing is labeled "Me 262 Modell I" the design still features the Bf 109E straight wing.

A close up photo of the AVA-Göttingen Me 262 Modell I wind tunnel model as seen from its nose port side. It appears that the diameter of the BMW gas turbines on this wind tunnel model more accurately reflect the fact that their diameter was indeed getting fatter due to the bigger compressor which was needed to increase the gas turbine's thrust.

Opposite:
Above: A port side view of AVA-Göttingen's Me 262 Modell I wind tunnel model. This model is evolving more and more into what would become the Me 262 A1-a series production version. Below: A poor quality photo of AVA-Göttingen's Me P-262 Modell I wind tunnel model as viewed from the front and above the model.

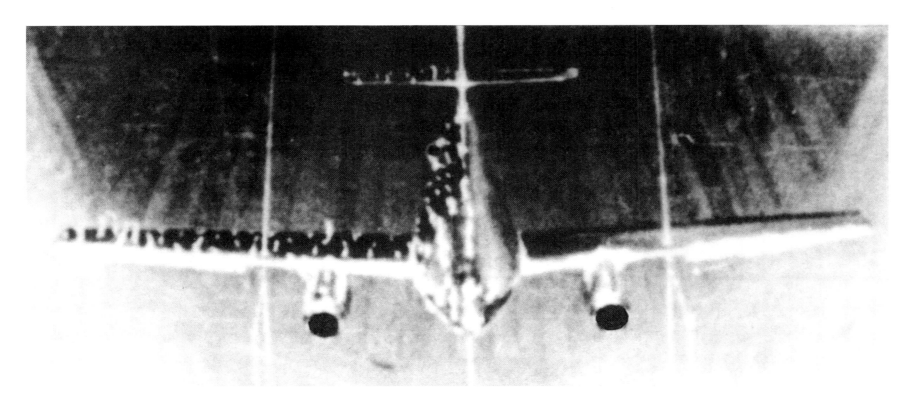

Messerschmitt Project Design 262 Evolving Post Modell I
Full 35° Sweep Low Wing AVA Wind Tunnel Model

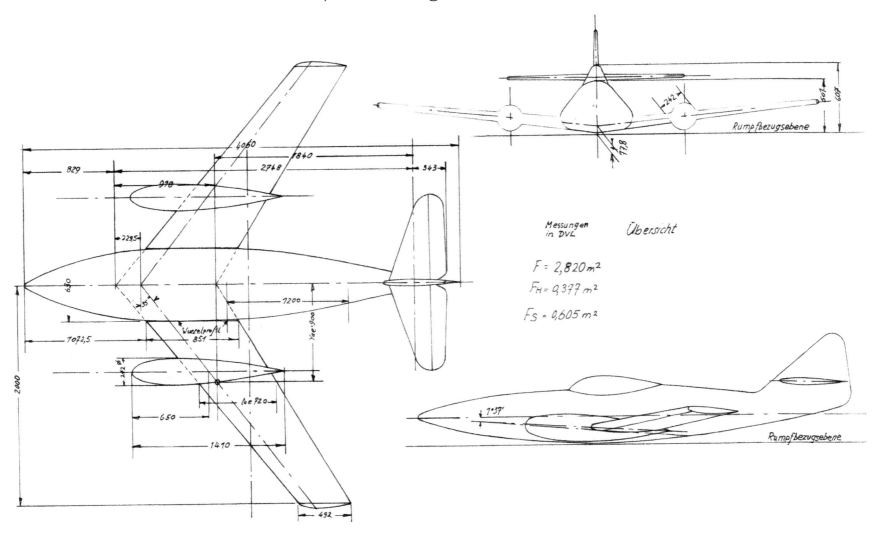

Messungen in DVL

Übersicht

$F = 2,820\,m^2$

$F_H = 0,377\,m^2$

$F_S = 0,605\,m^2$

With gas turbine locations settled on (beneath the wing) the Voigt Me 262 design team found that they required more change in the center of gravity. A full wing sweep back was determined and the team arrived at a 35 degree sweep back in order to bring the Me 262's airframe into balance. This poor quality 3-view drawing is from AVA-Göttingen and features a 35 degree wing sweep although AVA still has the BMW gas turbines going through the wing.

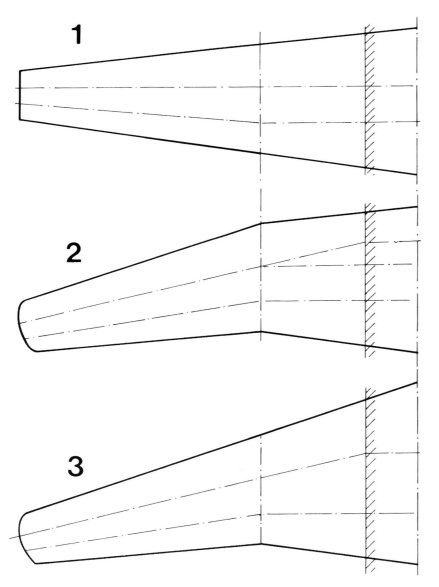

1

2

3

Voigt's design team started out with the idea of placing (1) a Bf 109E straight wing on their Me P-65/Me 262, then their ideas evolved to (2) a straight inner wing with an18 degree sweep on the outer wing, and finally (3), a full sweep back of 35 degrees for the entire wing as this pen and ink drawing illustrates.

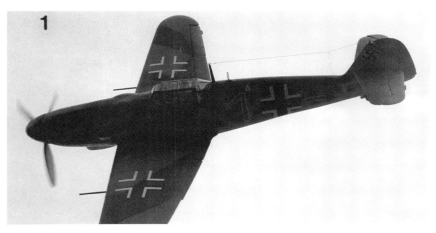

The Bf 109E and its straight wing (leading edge)...the style initially considered for the Me 262. Scale model and photograph by Günter Sengfelder.

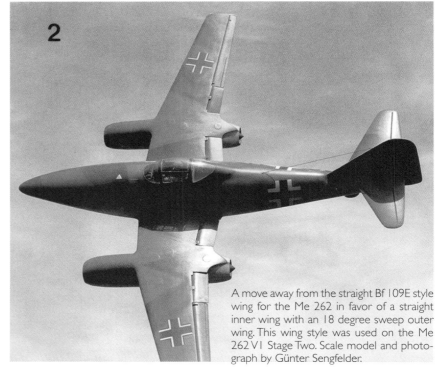

A move away from the straight Bf 109E style wing for the Me 262 in favor of a straight inner wing with an 18 degree sweep outer wing. This wing style was used on the Me 262 V1 Stage Two. Scale model and photograph by Günter Sengfelder.

3

The Me 262's final wing with its full 35 degree sweep back. Scale model and photograph by Günter Sengfelder.

Messerschmitt Project Design 262
Featuring (for the first time) a Nose Wheel

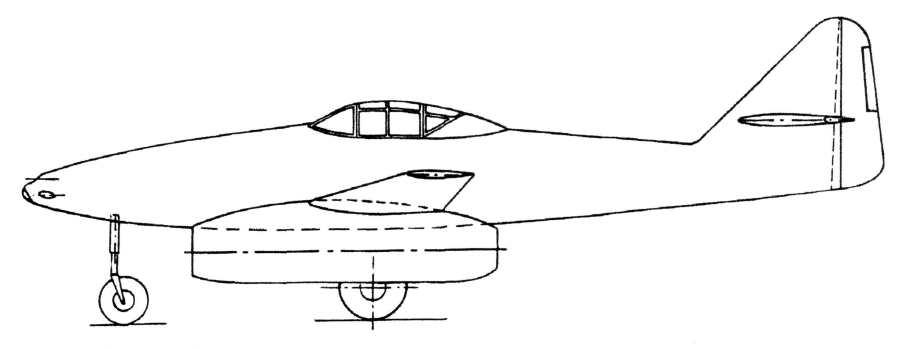

Port side profile of the evolving Me 262 design from November 1942, featuring a nose wheel design.

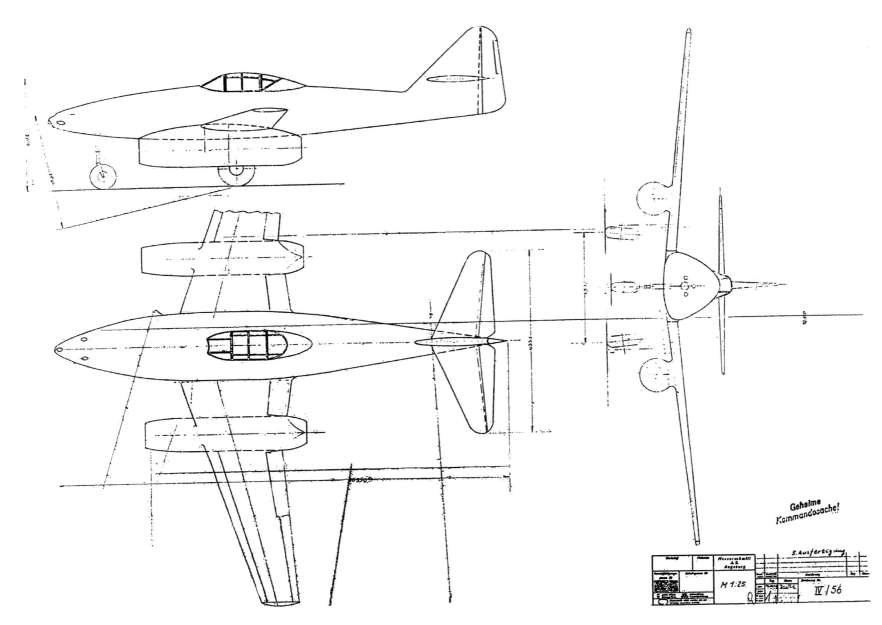

Gehsime
Kommandosache!

As Voigt's design team came closer and closer to their design objectives, this Me 262 design still was a tail dragger. Here is an early 3-view drawing of the Me 262 design from November 1942, with a crude nose wheel but a nose wheel nonetheless.

Messerschmitt Project Design 1070
Swept Back Low Wing Oval Fuselage Nose Wheel, BMW P-3302 Gas Turbines through the Wing

Parallel to the design work on Messerschmitt AG's "pursuit fighter" known as Project 1065, a second "pursuit fighter" design project was under consideration. Around Messerschmitt AG this project was known as P-1070 and it would be powered by two BMW P-3304 gas turbines. Its size overall was somewhat smaller in size to Me P-1 065 and the design team was directed by Messerschmitt AG's Project Bureau chief Dipl.-Ing. Hans Hornung (1910-1978). The major directive/purpose of P-1070 was a determined effort to come up with a design equal to performance and speed of the larger P-1065 but with a considerable reduction in weight. Hornung's team hoped to achieve this weight reduction through scaling back the P-1070's overall dimensions and wing surface area.

Messerschmitt AG's P-1070 design work was abandoned in 1940 in favor of the P-1065. By and large the P-1070's overall characteristics are unclear, but what is known includes the following:

- Wing span - 27 feet 3 inches [8.30 meters];
- Wing sweptback - 32 degrees at its leading edge;
- Wing area - 139.9 feet squared [13 meter squared];
- Wing loading (maximum) - 50 pounds per square [246 kilograms per square meter];
- Aspect ratio - 5.3 to 1;
- Length, overall - 26 feet 6 inches [8.10 meters];
- Height, overall - 9 feet 6 inches [2.90 meters];
- Weight, loaded - 6,172 to 7,054 pounds [2,800 to 3,200 kilograms];
- Speed, maximum anticipated - 683 miles/hour [1,100 kilometers/hour];
- Armament - two Mauser MK 151 20mm machine cannon mounted in the nose plus a single Rheinmetall-Borsig MK 131 13mm machine cannon positioned above the port side nose wheel well.

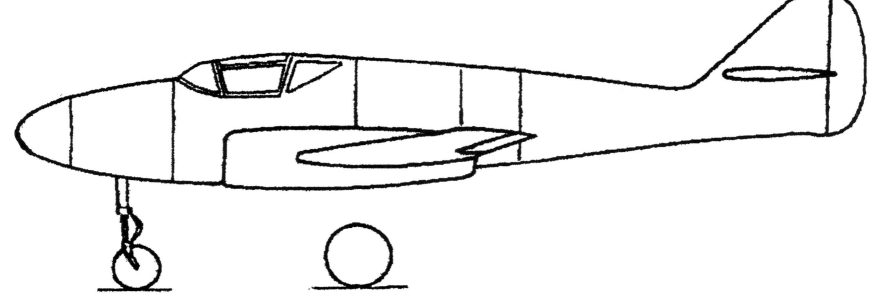

Port side illustration of the Me P-1070 "pursuit fighter."

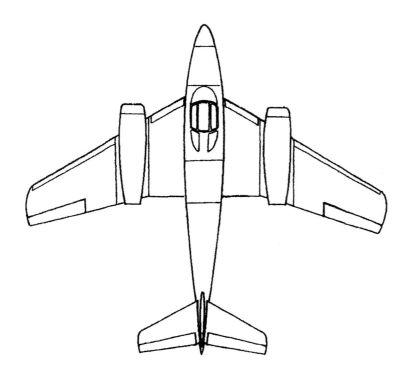

Willy Messerschmitt believed that his Me P-1070, when powered by twin BMW P-3304 gas turbines, would be capable of a top speed of 683.5 miles/hour [1,100 kilometers/hour]. Digital image by Ronnie Olsthoorn.

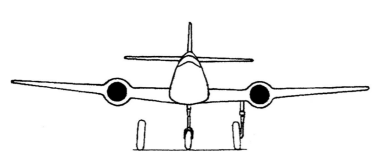

In 1940, Me P-1070 was a parallel design project to Voigt's Me P-1065...perhaps an alternate. Although Me P-1070 was a bit smaller flying machine, the design featured a tricycle landing gear, twin BMW P-3304 gas turbines which were to be placed inside the wing.

Like its design competitor the Me P-1065, the Me P-1070 initially was designed to have its twin gas turbines mounted in the wing. Digital image by Ronnie Olsthoorn.

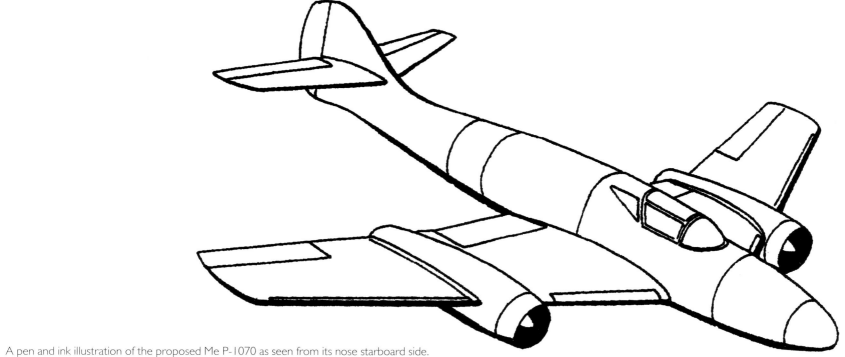

A pen and ink illustration of the proposed Me P-1070 as seen from its nose starboard side.

Left: The Me P-1070 was designed with a wing span of 27.23 feet [8.3 meters]. Digital image by Ronnie Olsthoorn. Below: The Me P-1070, a "pursuit fighter," would have been equipped with twin Mauser MG 151 15mm or 20mm machine cannon and a single Rheinmetall-Borsig MG 17 7.9mm machine cannon in the nose. Digital image by Ronnie Olsthoorn.

Me 262 Versuchs

(Experimental Versions)

On July 25th 1941, the RLM ordered Messerschmitt AG-Augsburg to construct five Me 262 prototypes and twenty pre-production versions. Construction began under the direction of Moritz Asam at Messerschmitt AG's Augsburg facilities. The RLM changed the number of research prototypes from time to time, however, 16 prototypes beyond Me 262 V1 stages one, two, and three would be constructed which include Me 262 V1-2 through Me 262 V-12.

- Me 262 V1 Stage One - Me 262 prototype. First flight April 18 th 1941, powered by a single Junkers Jumo 210G 640 horsepower piston engine. Flown 47 times for a total of 20 hours by flugkapitän Fritz Wendel and other Messerschmitt AG factory test pilots.
- Me 262 V1 Stage Two - In addition to the Junkers Jumo 210G piston engine, the fuselage fitted with two BMW P-3302 axial flow gas turbines. First and only flight with three engines was on March 25th 1942, by flugkapitän Fritz Wendel. Both BMW gas turbines failed shortly after take off and Wendel achieved the impossible by safely landing the heavy aircraft on piston power. This prototype was placed in Messerschmitt AG's hangar at Lager Lechfeid and never flown again.
- Me 262 V1 Stage Three - New fuselage to be powered by BMW gas turbines but redesigned to carry Junkers Jumo 004 gas turbines. First Me 262 prototype to achieve powered flight solely gas turbines. This achievement occurred on July 18th 1942, by Messerschmitt factory test pilot flugkapitän Fritz Wendel. The flight went well, however, its take off run required three-quarters of the 5,249 foot [1,600 meter] long Lager Lechfeld runway.
- Me 262 V1-2 - A mystery Me 262 prototype of which very little is known. It is known, however, to have been used to test the reduction of efficiency when using a long intake air duct for the Me P-1101. Used also for testing towed bombs and fuel tanks.
- Me 262 V2 - Another Me 262 mystery prototype making its first flight on October 2nd 1942. Appears to have been used to directional stability problems. Several surviving photographs show the aircraft modified with a large dorsal fin.
- Me 262 V2-2 [V056] - used to correct directional stability problems at high speed and tested with shortened vertical stabilizer of differing height. Used extensively to determine the change in aerodynamics over the tail assembly due to the installation of a four dipole radar antenna on the fuselage nose.
- Me 262 V3 - involved in high speed test flights to determine flight dynamics. Reached an unofficial speed record of 596.5 miles/hour on September 20th 1943. Test pilot reported experiencing severe vibrations and general instability.
- Me 262 V4 - first Me 262 air frame used to determine the flying machine's ability to carry external bombs.
- Me 262 V5 - experimented with the use of leading edge wing slots. Also used to test the suitability of Rheinmetall-Borsig take off rockets.
- Me 262 V5-2 [V167] - used to test modified rudder effectiveness, participated in bomb accuracy release trials when fitted with Wikingerschiff and ETC bomb racks along with Me 262 V10, flown over 300 times in extensive tests to improve directional stability.
- Me 262 V6 - first Me 262 with hydraulic activated landing gear. Its first flight occurred on October 17th 1943 Used also to determine the effect of extending its landing gear at varying speeds and whether the landing gear could be used as an air brake.
- Me 262 V7 - test prototype for seeking to work out the bugs involving the four Rheinmetall-Borsig MK 108 30mm cannon. Experimented with a new around view cockpit canopy, the prototype which later came to be used on all series production Me 262s.
- Me 262 V7-2 [V303] - used for flight performance measurements and also with heavy loads of external bombs for the "Snellbomber Program." First flight occurred on September 22nd 1944. Experimented with an all wood tail assembly, wheel brakes, and cabin ventilation.

- Me 262 V8 - used extensively flight testing the four nose mounted Rhein-metall-Borsig MK 108 30mm machine cannon seeking to perfect their operation under combat conditions. Completed 258 test flights with its first flight on March 18th 1944.
- Me 262 V8-2 [V484] - a highly modified Me 262 into a blitzbomber with the bombardier lying in a glazed nose cone dropping bombs with the TSA 2D bomb aiming device. First flight occurred in September 1944.
- Me 262 V9 - used to test modified control surfaces, experiments with swept back horizontal stabilizer, enlarge vertical stabilizer, and the "Rennkabine" racing cockpit canopy. First flight occurred on January 19th 1944.
- Me 262 V10 - used to test armor plating for pilot, Wikingerschiff bomb racks, bomb release mechanisms, participated with Me 262 V5-2 bomb release accuracy trials. Test the use of Rheinmetall-Borsig R-502 take off assist rockets. First flight occurred on January 5th 1944.
- Me 262 V1 1 [V555] - another blitzbomber with the bombardier lying prone behind a glazed nose cone operating the Lofte 7H gyro-stabilizing bomb sight. First flight occurred in February 1945.
- Me 262 V12 [V074] -tested as a prototype for the "Heimatschützer II" or home defense program with BMW 003R combined gas turbine and bi-fuel liquid rocket engine. Static ground tests were made, once one of the BMW 003R engines exploded. No flight tests are known to have occurred.

Me 262 V1 Stage 1

- The Me 262 V1 Stage 1 is also known as the Me 262 V1-1.
- Werknummer assigned was 262 000 01.
- First, the prototype was known as Me P-1065. Later it was redesignated by the RLM as 8-262.
- Me 262 V1 Stage 1 was assigned Stammkennzeihen or radio call code PC+UA.
- Me 262 V1 Stage 1 was designed for two BMW P-3302 axial flow gas turbine engines with a factory rated static thrust at sea level of 1,000 pounds, however, it is doubtful that they reached anymore than 75% of their rated output.
- No lettering or numerals appeared on the Me 262 V1 Stage 1's fuselage.
- Me 262 V1 Stage 1 was first flight tested/flown powered solely by its single Junkers Jumo 210G (680 horsepower) piston engine. This was because its intended twin BMW P-3302 axial gas turbine engines were not certified flight ready for field testing. Thus the gas turbines were not shipped to Messerschmitt AG.
- The maiden flight of Me 262 V1 Stage 1 with its single Jumo 210G piston engine occurred on April 18th 1941.
- Messerschmitt AG factory test pilot flugkapitän Fritz Wendel made the first flight. Other test pilots test flew Me 262 V1 Stage 1 afterward.
- During its flight testing, Me 262 V1 Stage 1's aluminum fuselage was left unpainted.
- On July 18th 1943, Me 262 V1 Stage 1 was flown with cockpit pressurization plus three Mauser MG 151 20mm machine cannon in the nose.
- Me 262 V1 Stage 1 was damaged from a USAAF bombing raid on July 7th 1944. It was not repaired and never flown again.
- Final disposition of Me 262 V1 Stage 1 at war's end in Europe on May 8th 1945, is unclear.

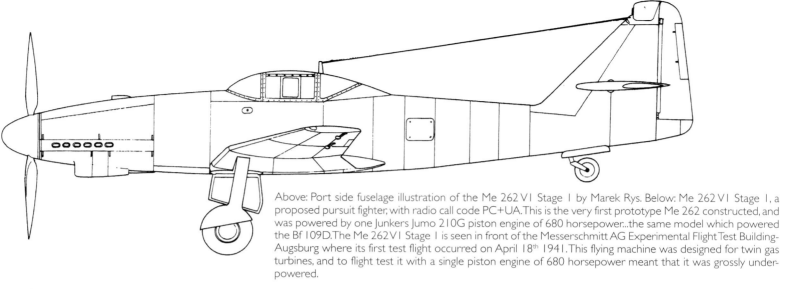

Above: Port side fuselage illustration of the Me 262 V1 Stage 1 by Marek Rys. Below: Me 262 V1 Stage 1, a proposed pursuit fighter, with radio call code PC+UA. This is the very first prototype Me 262 constructed, and was powered by one Junkers Jumo 210G piston engine of 680 horsepower...the same model which powered the Bf 109D. The Me 262 V1 Stage 1 is seen in front of the Messerschmitt AG Experimental Flight Test Building-Augsburg where its first test flight occurred on April 18th 1941. This flying machine was designed for twin gas turbines, and to flight test it with a single piston engine of 680 horsepower meant that it was grossly under-powered.

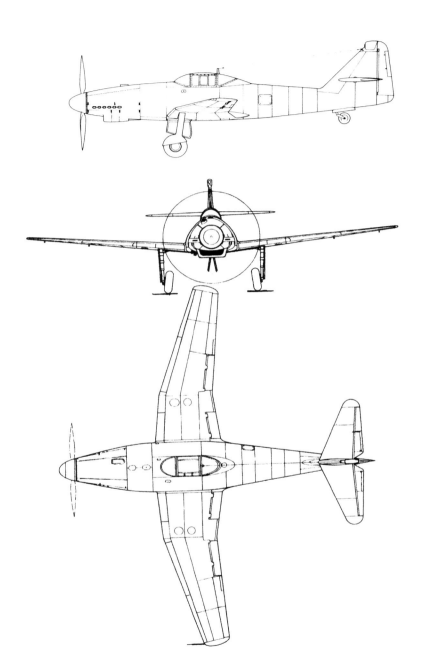

Left: Me 262 V1 Stage 1 three-view line drawing by Marek Rys. Above: Junkers Jumo 210G inverted "V" piston engine of 680 horsepower, which powered the Me 262 V1 Stage 1 prototype, as seen from its from starboard side. Notice the square box-like exhaust port openings near the bottom. Below; A Junkers Jumo 210G inverted "V" piston engine as seen from below. The engine's front is to the left in the photograph.

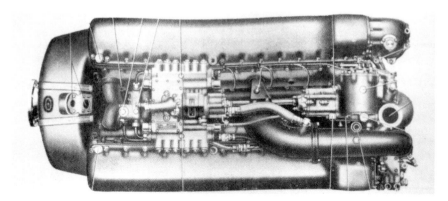

A typical installation of the Junkers Jumo 210G in a Bf 109D fighter, however, its placement in the Me 262 V1 Stage 1 prototype would have appeared identical right down to the exhaust ports and large radiator below the engine's rocker arm covers.

The Me 262 V1 Stage 1 prototype PC+UA seen from its starboard side with its cockpit canopy open to starboard, too. The scoop on the nose aft the two bladed propeller is the Junkers Jumo 210G's air intake. Parked out in front of the prototype is a Caudron C-445.

A close up of the Me 262 V1 Stage 1 prototype as seen from its port side featuring its cockpit wind screen and full canopy. Notice that the cockpit canopy is made up of box like panels. Below near the port wing root appears to be a pilot's parachute pack.

Flugkapitän Fritz Wendel...the only known test pilot of the Me 262 V1 Stage 1 prototype seen here wearing a leather flight jacket. He reported that the maximum forward speed with the Junkers Jumo 210G in level flight was only 260 miles/hour [418 kilometers/hour] and that its climbing ability was extremely poor. However, Wendel stated that the flight characteristics of the airframe were excellent, good all around view from the cockpit, and excellent flight controls. During dives when speeds of 336miles/hour [540 kilometers/hour] were achieved, Wendel reported experiencing slight elevator vibrations. Overall, Wendel would test fly the Me 262 V1 Stage 1 47 times before the BMW gas turbines were installed.

The Me 262 V1 Stage 1 prototype featuring its nose/engine as seen from the port side. The rectangular openings at the top of the photo are the Junkers Jumo 210G's exhaust ports. Below, in the center of the photo, is the 210G's radiator opening.

The Me 262 V1 Stage 1 and its huge radiator needed to cool the 680 horsepower Junkers Jumo 210G piston engine. The metal ladder out in front the port main wheel is required to get up on the wing, do maintenance on the engine, or enter/exit the cockpit.

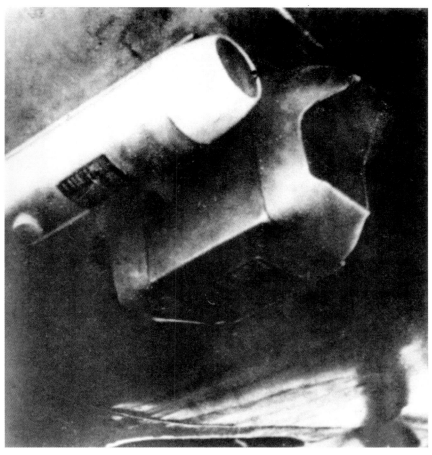

Two starboard side Junkers Jumo 210G engine related items from the Me 262 V1 Stage 1 appear in this photo. First, is the exit end of the 210G's radiator duct. Second, is the 210G's oil cooler which was placed parallel to aft of the radiator.

Me 262 V1 Stage 2

- The Me 262 V1 Stage 2. This version is also known as Me 262 V1-2.
- Me 262 V1 Stage 2's radio call code continued to be the same as Me 262 V1 Stage 1 ... PC+UA.
- Me 262 V1 Stage 2 was fitted with twin BMW P-3302 axial flow gas turbines sometime between September and December 1941. The exact date is unclear. Its single Junkers Jumo 21OG aero piston engine was retained for safety reasons thus making the Me 262 V1 Stage 2 a hybrid power plant flying machine ... one engine of old technology and two new.
- On March 25th 1942, Me 262 V1 - Stage 1, with 47 previous test flights behind it for a total of 20 hours, lifted off the tarmac at Lager Lechfeld as Me 262 V1 Stage 2. Flugkapitän Fritz Wendel radioed that he had "one (engine) turning and two (engines) burning." Then shortly after lift off both BMW P-3302 gas turbines suffered compressor blade failure. With heavy dark smoke billowing from both gas turbines ... they stalled, then froze. Wendel, through sheer suburb piloting skills struggled to land the heavy hybrid equipped flying machine with three engines ... only through the underpowered Junker Jumo 21OG piston engine. Wendel avoided a certain crash landing for a pilot with lesser skills.
- The Me 262 V1 Stage 2's aluminum fuselage was not painted, but flown in its one time flight in its aluminum finish, as had the Me 262 V1 Stage 1 before.
- Final disposition at war's end in Europe on May 8th 1945, is unclear.

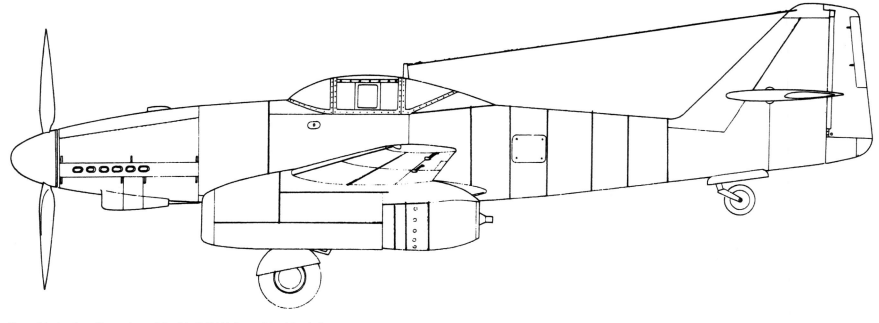

Port side fuselage illustration of the Me 262 V1 Stage 2 by Marek Rys.

The Me 262 VI Stage 2 as seen from its nose starboard side. Its two bladed propeller and propeller spinner were painted a dark green. The same, too, for the metal frame work of the wind screen and cockpit canopy. Digital image by Mario Merino.

The Me 262 VI Stage 2 as seen from below featuring its port side. Notice its large radiator, which was needed for its 680 horsepower Junkers Juno 210G piston engine. Digital image by Mario Merino.

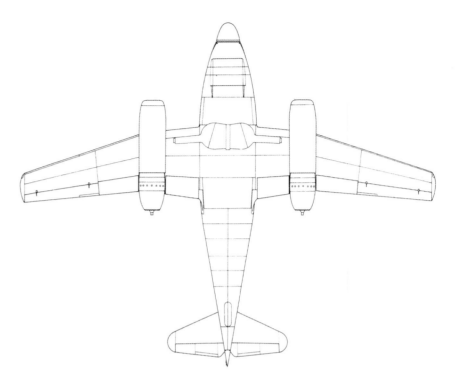

An underside view of the Me 262 V1 Stage 2 illustrated by Marek Rys.

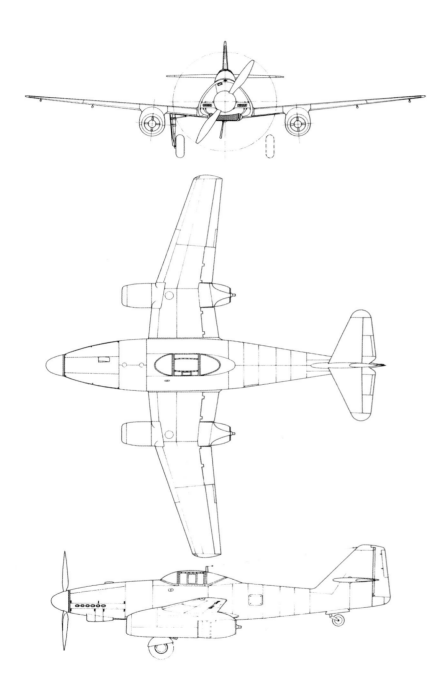

Me 262 V1 Stage 2 three view line drawing by Marek Rys.

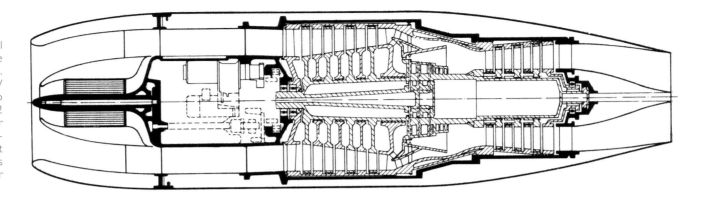

An illustration of the BMW P-3304 axial flow gas turbine which helped power the Me 262 V1 Stage 2 on March 25th 1942. This gas turbine was also known as BMW 002. BMW engineers were seeking to develop a gas turbine such as the 002 which would be smaller and lighter for the same amount of thrust as the Junkers Jumo 004 gas turbine. Notice that BMW's engineers had placed all the gas turbine's accessories equipment in the air intake.

One of the few known photos of the hybrid engined Me 262 V1 Stage 2 stored in a hangar at Messerschmitt AG's Lager Lechfeld after its first and only test flight on March 25th 1942. It appears that the BMW P-3304 gas turbines have been removed from their cowlings...perhaps taken back to BMW-Berlin/Spandau and Dr.-Ing. Hermann Oestrch to determine the cause of both engine compressors failing at the same time shortly after lift off.

The cockpit canopy of the Me 262 VI Stage 2 with its three glass panels on the port side. The open canopy, which opened to starboard, is seen from its rear starboard side.

Messerschmitt Bf 109TL

About mid-1942 the RLM feared that Messerschmitt AG's Me 262 "pursuit fighter" might not overcome its initial difficulties any time soon. The RLM believed that a parallel gas turbine powered project based on the aging Bf 109 piston powered fighter might make a lot of sense. It would be designated Bf 109 TL ... TL meaning gas turbine. For a while, Messerschmitt AG designers were seriously considering hanging two Junkers Jumo 004 gas turbines from the wing of a Bf 109 and doing it as fast as possible. As an interim solution, the plan was to take the fuselage of a Me 155, which was a fighter-bomber based on the Bf 109, the tricycle landing gear of the Me 309, and the wings of the planned Me 409, and combine them all into a twin gas turbine powered pursuit fighter.

Rumors suggest that at least one Bf 109 TL was produced, but this may be just a matter of belief rather than a matter of fact. Nevertheless, it appears that the idea of a gas turbine powered Bf 109 was canceled shortly after the successful flight of the Me 262 V1 Stage Three.

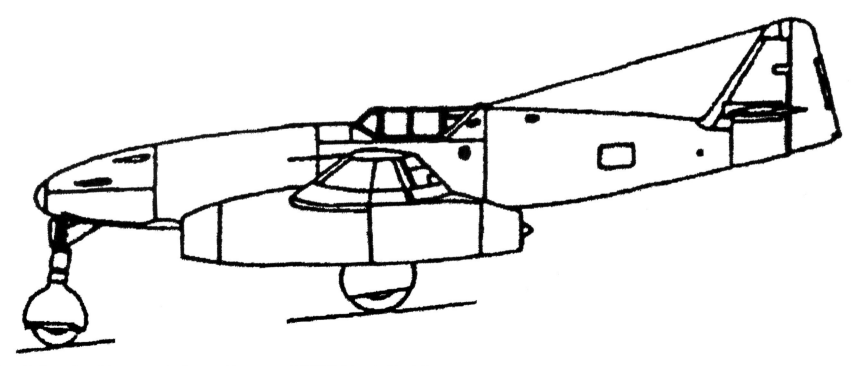

Port side illustration of the proposed twin gas turbine powered Bf 109TL fighter by Marek Rys.

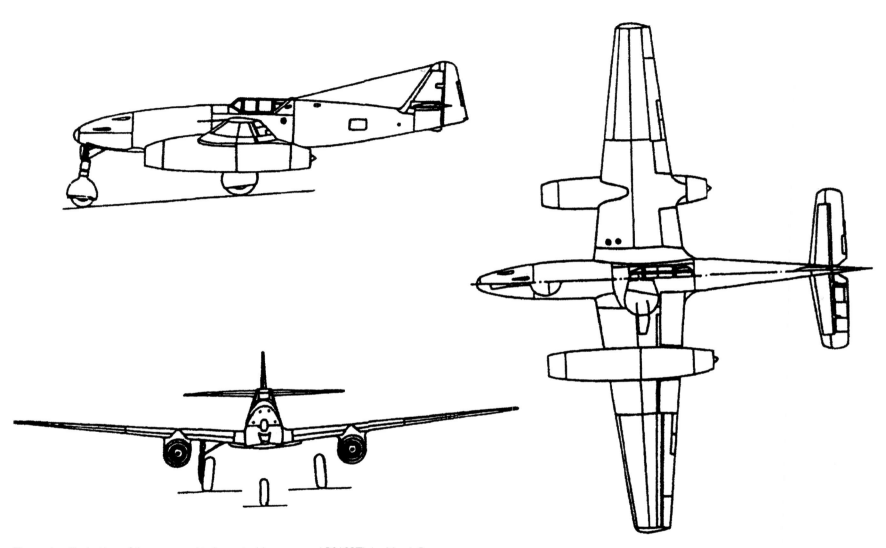

Three view illustration of the proposed twin gas turbine powered Bf 109TL by Marek Rys.

Me 262 V1 Stage 3

- The Me 262 V1 Stage 3 is also known as Me 262 V1 -3.
- The version known as Stage 3 was originally designed for two BMW P-3302 gas turbine engines. However, BMW-Berlin/Spandau abandoned their P-3302 and P-3304 gas turbine projects in late March 1943. They found them unable to meet their design thrust rating without a complete redesign. With this news Messerschmitt AG was forced to use the gas turbines built by Junkers Jumo. Before doing so, Messerschmitt AG had to rebuild the Stage 3's wing and fuselage in order to mount the them on their prototype flying machine. The required modifications did not seem to be too difficult to the Messerschmitt AG engineers. The Junkers Jumo 210G piston engine would be removed during the modifications convincing proof of the much greater confidence the RLM and Messerschmitt AG had in the Junkers Jumo 004A-O gas turbines.
- Each Junkers Jumo 004A-O was rated at 1,850 pounds of static thrust while the BMW P-3302 was been overrated at 1,000 pounds static thrust.

- First flight of the Me 262 V1 Stage 3 was on July 18th 1942 from the air base at Leipheim near Augsburg. Messerschmitt AG factory test pilot flugkapitän Fritz Wendel.
- Due to the high reliability and success of the Junkers Jumo 004A-0 gas turbines verses the BMW P-3302, the Jumo 004 because the gas turbine of choice and remained so right up to war's end in Europe on May 8th 1945.
- Stage 3's fuselage was painted in RLM 74, 75, and 76.
- Stage 3 was equipped with three Mauser MG 151 20mm machine cannon placed in the nose ... similar to the arrangement proposed in the Me P-1065 design.
- Me 262 V1 Stage 3, after its initial test flights, was used to test "Baldrian" airborne acoustic homing devices. Originally "Baldrian" was a code word for acoustic homing devices for torpedoes manufactured by Telefunken.
- Stage 3 suffered a crash landing on July 7th 1944, after having made 95 test flights. At the time it was being piloted by test pilot Heinz Herlitzius. The flying machine's port main wheel collapsed while Herlitzius was attempting to correct a swerve upon landing.
- The Me 262 V1 Stage 3 was not repaired but written off as unrepairable.

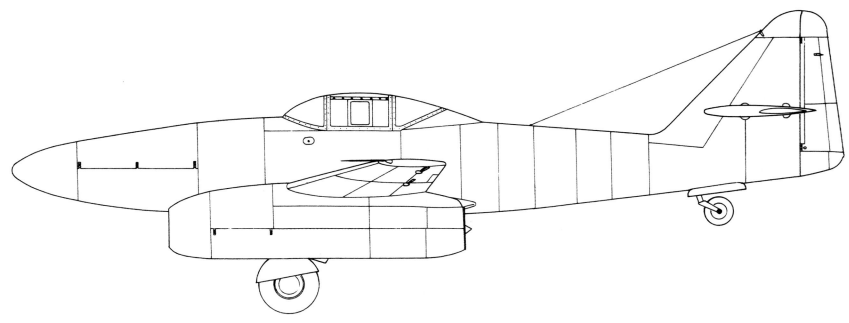

Port side line drawing of the Me 262 V1 Stage 3 (PC+UB) by Marek Rys.

The Me 262 V1 Stage 3 with its radio call code PC+UB carried over from Me 262 V1 Stage 2. This revolutionary flying machine was photographed sitting outside with its cockpit canopy covered over with a tarp to protect it from the elements as it waits for two Junkers Jumo 004A-O gas turbines to be delivered.

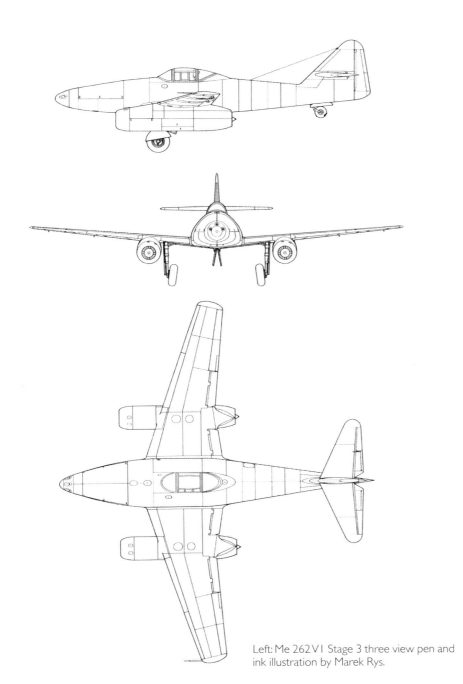

Left: Me 262 VI Stage 3 three view pen and ink illustration by Marek Rys.

A Messerschmitt AG gas turbine specialist is seen lifting the cowlings off an early Me 262 (VI+AA) starboard Junkers Jumo 004A-O gas turbine.

Typical of Me 262 VI Stage 3's starboard side Junkers Jumo 004A-O gas turbine. This photo is of VI+AA. Notice that the Riedel gas turbine starter cone has a series of perforations not found in the Jumo 004B gas turbines...perhaps to allow more air flow to the starter motor.

The Me 262 VI Stage 3 (PC+UB) seen here fitted with its twin Junkers Jumo 004A-O gas turbines. Junkers Jumo delivered two of their gas turbines to Messerschmitt AG on June 1st 1942, and the flying machine's first flight was six weeks later on July 18th 1942. Notice that now the fuselage and tail assembly have been given a new paint job from its pre- Junkers Jumo days.

A direct port side view of what would be typical on the Me 262 VI Stage 3's Junkers Jumo 004A-O gas turbine. This photo is of VI+AA. The gas turbine's air intake to is the left in the photo followed by its air compressor, combustion chambers and thrust nozzle.

The Me 262 V1 Stage 3. The metal bulk head with its several round cutouts signifies that this flying machine was powered by Junkers Jumo 004s because the previous BMW P-3302 did not have this type of bulkhead.

A close up view of the extremely rare Junkers Jumo 004A-O gas turbine which powered the Me 262 V1 Stage 3 as seen from its port side and featuring its air compressor section. To the far left in the photo but not seen is its Riedel RLM 9-7034A gasoline starter engine.

The Junkers Jumo 004A-O gas turbine as seen from its port side with its Riedel RLM 9-7034A starter motor to the far left on the photo.

The Junkers Jumo 004 A-O gas turbine as seen from its rear port side and featuring the combustion chamber section.

Left: Me 262 test pilot Feldwebel (Sargent) Heinz Herlitzius. On June 23rd 1944, while piloting Me 262 "S-2" achieved 624 miles/hour [1,004 kilometers/hour] in a 35 degree dive from 22,970 feet [7,000 meters] altitude. Below: Port side fuselage illustration of the Me 262 V1 Stage 3 modified for static cockpit pressurization tests. Drawing by Marek Rys.

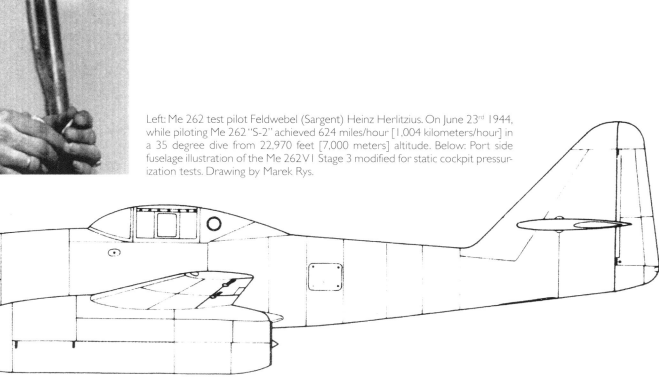

The Me 262 V1 Stage 3 in Messerschmitt AG's hangar and no longer being airworthy, however while stored in the hangar it was used for cockpit pressurization tests. The way to tell is that when it was being used for cockpit pressurization, Messerschmitt AG engineers removed the aft wrap around glass and covered it over with aluminum. They then installed a porthole through the aluminum. This porthole can be seen in this photograph.

The Me 262 V1 Stage 3 seen in the upper left corner of the photo. Lechfeld had just suffered a bombing raid by the USAAF (July 18th 1944) and there is rumble throughout hangar #5. However, it appears that Me 262 V1 Stage 3 survived the bombing raid with very little physical damage while all around it there is considerable destruction.

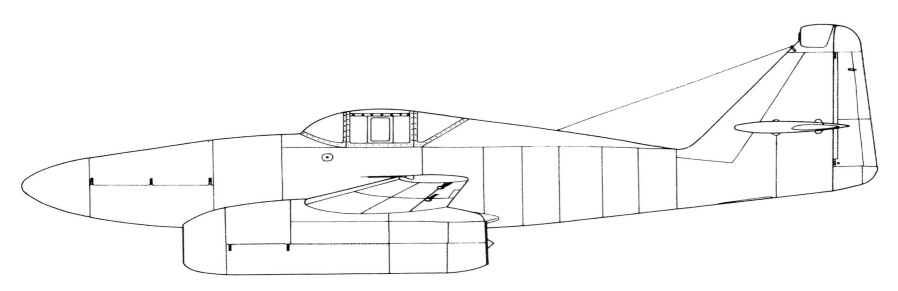

Port side fuselage illustration of the modified cockpit on the Me 262 V1 Stage 3. Drawing by Marek Rys.

Me 262 Versuchs (Experimental Versions)

On July 25th 1941, the RLM ordered Messerschmitt AG-Augsburg to construct five Me 262 prototypes and twenty pre-production versions. Constructed began under the direction of Moritz Asam at Messerschmitt AG's Augsburg facilities. The RLM changed the number of research prototypes from time to time, however, 16 prototypes beyond Me 262 V1 stages one, two, and three would be constructed which include Me 262 V1-2 through Me 262 V12.

- Me 262 V1 Stage One - Me 262 prototype. First flight April 18th 1941, powered by a single Junkers Jumo 210G 640 horsepower piston engine. Flown 47 times for a total of 20 hours by Flugkapitän Fritz Wendel and other Messerschmitt AG factory test pilots.
- Me 262 V1 Stage Two - In addition to the Junkers Jumo 210G piston engine, the fuselage fitted with two BMW P-3302 axial flow gas turbines. First and only flight with three engines was on March 25th 1942, by Flugkapitän Fritz Wendel. Both BMW gas turbines failed shortly after take off and Wendel achieved the impossible by safely landing the heavy aircraft on piston power. This prototype was placed in Messerschmitt AG's hangar at Lager Lechfeld flugplatz and never flown again.
- Me 262 V1 Stage Three - New fuselage to be powered by BMW gas turbines but redesigned to carry Junkers Jumo 004 gas turbines. First Me 262 prototype to achieve powered flight solely gas turbines. This achievement occurred on July 18th 1942, by Messerschmitt factory test pilot flugkapitän Fritz Wendel. The flight went well, however, its take off run required three-quarters of the 5,249 foot [1,600 meter] Lager Lechfeld runway.
- Me 262 V1-2 - A mystery Me 262 prototype of which very little is known. Known to have been used to test the reduction of efficiency when using a long intake air duct for the Me P-1101. Used also for testing towed bombs and fuel tanks.
- Me 262 V2 - Another Me 262 mystery prototype making its first flight on October 2nd 1942. Appears to have been used in an attempt to solve directional stability problems. Several surviving photographs show the aircraft modified with a large dorsal fin.

- Me 262 V2-2 [V056] - used to correct directional stability problems at high speed and tested with shortened vertical stabilizer of differing height. Used extensively to determine the change in aerodynamics over the tail assembly due to the installation of a four dipole radar antenna on the fuselage nose.
- Me 262 V3 - involved in high speed test flights to determine flight dynamics. Reached an unofficial speed record of 596.5 miles/hour on September 20th 1943. Test pilot reported experiencing severe vibrations and general instability.
- Me 262 V4 - first Me 262 air frame used to determine the flying machine's ability to carry external bombs.
- Me 262 V5 - experimented with the use of leading edge wing slots. Used also to test the suitability of Rheinmetall-Borsig take off rockets.
- Me 262 V5-2 [V167] - used to test modified rudder effectiveness, participated in bomb accuracy release trials when fitted with Wikingerschiff and ETC bomb racks along with Me 262 V10, flown over 300 times in extensive tests to improve directional stability.
- Me 262 V6 - first Me 262 with hydraulic activated landing gear. Its first flight occurred on October 17th 1943. Used also to determine the effect of extending its landing gear at varying speeds, and whether the landing gear could be used as an air brake.
- Me 262 V7 - test prototype for seeking to work out the bugs involving the four Rheinmetall-Borsig MK 108 30mm cannon. Experimented with a new around view cockpit canopy, the prototype which later came to be used on all series production Me 262s.
- Me 262 V7-2 [V303] - used for flight performance measurements and also with heavy loads of external bombs for the "Snellbomber Program." First flight occurred on September 22nd 1944. Experimented with an all wood tail assembly, and improved wheel brakes and cabin ventilation.
- Me 262 V8 - used extensively flight testing the four nose mounted Rheinmetall-Borsig MK 108 30mm machine cannon seeking to perfect their operation under combat conditions. Completed 258 test flights with its first flight on March 18th 1944.

- Me 262 V8-2 [V484] - a highly modified Me 262A-1a into a blitzbomber with the bombardier lying in a glazed nose cone dropping bombs with the TSA 2D bomb aiming device. First flight occurred in September 1944.
- Me 262 V9 - used to test modified control surfaces, experiments with swept back horizontal stabilizer, enlarge vertical stabilizer, and the "Rennkabine" racing cockpit canopy. First flight occurred on January 19th 1944.
- Me 262 V10 - used to test armor plating for pilot, Wikingerschiff bomb racks, bomb release mechanisms, participated with Me 262 V5-2 bomb release accuracy trials. Test the use of Rheinmetall-Borsig R-502 take off assist rockets. First flight occurred on January 5th 1944.
- Me 262 V1 1 [V555] - another blitzbomber version with the bombardier lying prone behind a glazed nose cone operating the Lofte 7H gyro-stabilizing bomb sight. First flight occurred in February 1945.
- Me 262 V12 [V074] -tested as a prototype for the "Heimatschützer II" or home defense program with BMW 003R combined gas turbine and bi-fuel liquid rocket engine. Static ground tests were made, once one of the BMW 003R engines exploded. No flight tests are known to have occurred.

Me 262 V1-1

- A mystery Me 262 flying machine designated Me 262 V1-1, about which very little is known and understood.
- Werknummer 130015.
- Date, location assembled, and placed into service is unclear.
- Radio call code is unclear.
- Only a few photographs are thought to exist of the Me 262 V1-1, usually showing its nose and tail assembly covered with a fishnet. The purpose/need for this covering is unclear.

- Me 262 V1-1 is known to have been used experimentally to test the loss of thrust resulting from the friction of a long air intake duct for gas turbine engines. The results of this research would be applied to the prototype Messerschmitt P-1101 with its single Heinkel-Hirth HeS 011 axial flow gas turbine engine.
- In addition to testing the loss of efficiency of long air intake ducts, Me 262 V1-1 is known to have been involved in testing towed bombs and fuel tanks.
- Final disposition of Me 262 V1-1 at war's end in Europe on May 8th 1945 is unclear.

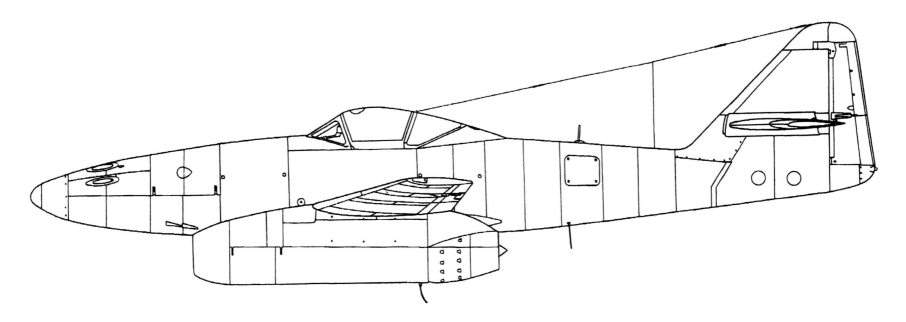

Me 262 V1-2

- The Me 262 V1 Stage 2. This version is also known as Me 262 V1 -2.
- Me 262 V1 Stage 2's radio call code continued to be the same as Me 262 V1 Stage 1 ... PC+UA.

- Me 262 V1 Stage 2 was fitted with twin BMW P-3302 axial flow gas turbines sometime between September and December 1941. The exact date is unclear. Its single Junkers Jumo 210G aero piston engine was

Me 262 V1-2 seen being tested with a 6 foot [2 meter] long air intake tube to measure the loss in thrust due to friction inside the tube. This research was in anticipation of the Messerschmitt Me P-1101 fighter with its long air intake leading to its single Heinkel-Hirth HeS 011 axial flow gas turbine.

The Messerschmitt Me P-1101 as seen from its port side and its single Heinkel-Hirth HeS 011 gas turbine. The gas turbine's placement inside the fuselage would have required a long air intake duct and the Me 262 V1-2 was being used to measure potential loss in thrust. Scale model and photograph by Günter Sengfelder.

retained for safety reasons thus making the Me 262 V1 Stage 2 a hybrid power plant flying machine ... one engine of old technology and two new.

- On March 25th 1942, Me 262 V1 - Stage 1 with 47 previous test flights behind it for a total of 20 hours, lifted on the tarmac at Lager Lechfeld as Me 262 V1 Stage 2. Flugkapitön Fritz Wendel radioed that he had "one (engine) turning and two (engines) burning." Then shortly after lift off both BMW P-3302 gas turbines suffered compressor blade fail-ure. With heavy dark smoke billowing from both gas turbines ... they stalled then froze. Wendel, through sheer suburb piloting skills struggled to land the heavy hybrid equipped flying machine with three engines ... only through the underpowered Junkes Jumo 210G piston engine. Wendel avoided a certain crash landing for a pilot with lesser skills.

- The Me 262 V1 Stage 2's aluminum fuselage was not painted, but flown in its one time flight in its aluminum finish, as had the Me 262 V1 Stage 1 before.

- Final disposition at war's end in Europe is unclear.

The Me 262 V1-2 on the right and seen with its ever present fishnet camouflage netting over its nose and tail plane assembly. To the Me 262 V1-2's left is the Me 262 A-2a/U2 fast bomber with Lofte 7H bombardier variant.

Beneath the port wing of a Messerschmitt Me 321 "Gigant" glider transport is the fully cloaked Me 262 V1-2...all covered over by its fishnet camouflaging. Notice that even its cockpit canopy is covered. Another "weissblau" colored Me 262 appears in the center of the photograph...perhaps one of the Me 262 "S" types.

Me 262 V2

This Me 262 "Versuchsflugzeug" flying machine, known simply as "Me 262 V2," is a mystery without many clues. In only two known photographs, this Me 262 is seen with a tricycle landing gear. It may have been one of the first five "V prototypes completed by Messerschmitt AG. This would have been from a contract awarded the company by the RLM on July 25th 1942. Also, on this date, Messerschmitt AG was given a contract to construct twenty pre-production "O-Serien," or simply "S" Me 262s. The first ten Me 262 "Versuchsflugzeug" and "S" prototypes were to be delivered between June and October 1942.

Werknummer of Me 262 V2 is not clear.

Radio call code of Me 262 V2 is not clear.

It appears from the few photographs which have survived that Me 262 V2 was involved at one time in attempts to improve the Me 262's known directional stability problems. Two photos of the Me 262 V2 show the placement of a crude opaque dorsal fin extending about 3 feet [1 meter] in height aft the nose cone and running along the fuselage surface perhaps to the windscreen ... perhaps even further aft beyond the windscreen. It is unclear. However, this dorsal fin is secured to the fuselage by metal vertical supports and guy wires. In all, the purpose of this seemingly makeshift dorsal fin research is unclear.

Final disposition of the Me 262 V2 is unclear.

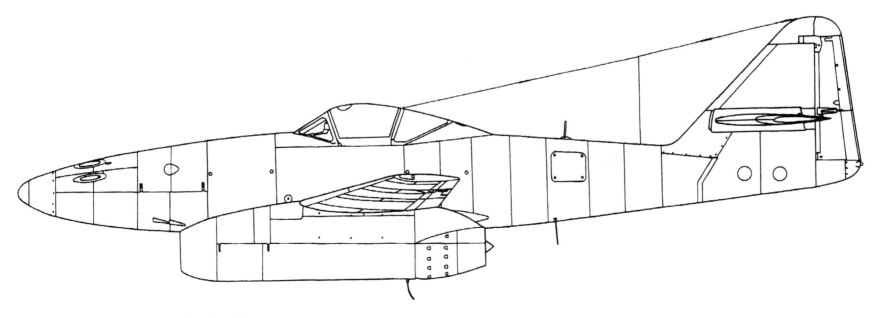

Port side fuselage profile illustration of the Me 262 V2. Drawing by Marek Rys.

The Me 262 V2 appearing with a vertical fin attached to the top of the fuselage. The purpose of this flimsily looking item is unclear. The material appears opaque and it is supported by vertical rods. It is also unclear how far aft the fuselage this modification extends.

A nose on view of Me 262 V2 and its vertical fin like modification. Notice that the fin itself is secured by guy wires leading down to the fuselage.

Me 262 V2-2 (2ⁿᵈ Version Known As [V056])

- Known as Me 262 V2-2.
- Werknummer 170056...hence the designation "V056," which was taken from the last three werknummer digits.
- Overall, the fuselage is a typical Me 262 A-1a airfame.
- V056 is perhaps the most thoroughly used Me 262 "Versuchs," or mule, used for all sorts of experimental testing. It was used extensively in experimenting with night fighting search radar and IFF wing mounted mast antennae.
- At one time V056 was fitted with an experimental Siemens Funk Gerät FuG 218 "Neptun V1 " airborne radar array. Its purpose was to test and evaluate the night fighter capability of the typical Me 262 A-1a air frame. The four dipole antenna was located on the V056's nose. It was in the form of an 'X'

but no other distinctive modifications appeared to have been made to this experimental flying machine. However, V056's tail assembly was painted black at one time when white tufts of cotton were attached to the black painted tail assembly. Its purpose was to observe the air stream around V056's tail assembly from the disruption in air flow from its "X" antenna array.

- Flying tests of the V056 were conducted beginning sometime in January 1945. It was flown by several highly experienced pilots including Heinz Herlitzius and night fighter ace Oberleutnant (First Lieutenant) Kurt Welter. Welter had achieved 56 aerial victories including 20 at night ... among them British heavy bombers and Mosquitos.
- V056 was found at war's end in Europe at Lager Lechfeld piled among other spent Me 262s.

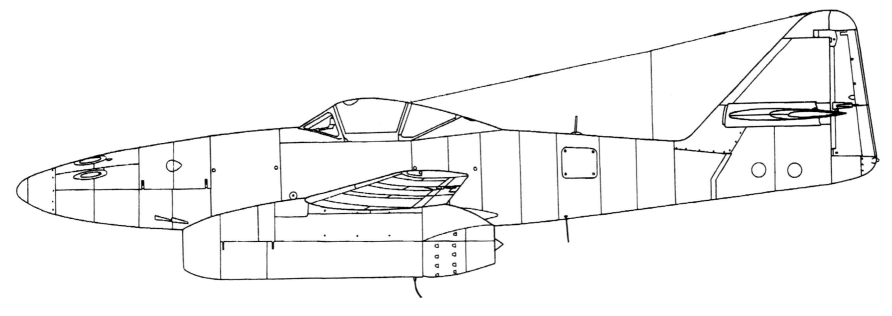

Port side fuselage profile illustration of the Me 262 V2-2. Drawing by Marek Rys.

A nose on view of the Me 262 V2 2nd version and known as V056. Werknummer is 170056. This V056 appears new and is being used for testing the effectiveness of only two Rheinmetall-Borsig MK 108 30mm machine cannon. Notice that the upper gun ports have been closed off.

A ground level nose starboard side view of the Me 262 V2 2nd version. Appearing factory fresh the machine has yet to have bomb racks installed for its intended use as a fighter-bomber prototype.

Me 262 V2-2 V056 was a test mule for the upcoming Me 262 A- 2a bomber. Initially, as shown here painted in RLM 76, the tests involved closing off the two upper gun ports and the V056 was used to test the effectiveness of firing with only two Rheinmetall-Borsig MK 108 30mm machine cannon.

The Me 262 V2 2nd version [V056]. The officer appears to be holding the spent shell shut for the aircraft's starboard side Rheinmetall- Borsig MK 108 30mm machine cannon.

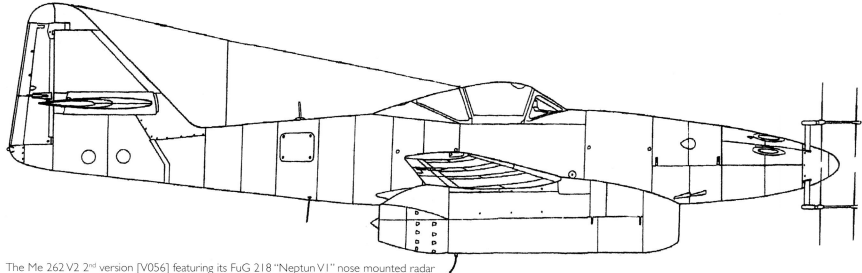

The Me 262 V2 2nd version [V056] featuring its FuG 218 "Neptun VI" nose mounted radar array. Line drawing by Marek Rys.

Among the electronics tested, [V056] appears most famous for those involving the "X" shaped antenna array for Siemens (Funk Gerät) FuG 218 "Neptun" search radar for an upcoming dedicated two man Me 262 B-1a/U1 night fighter. The V056 is seen from its starboard side.

This Me 262 V2 2nd edition appears well used and not very clean...very unlike many pristine Me 262s found in aviation museums throughout the world. The tail assembly has been painted black (RLM 22). During its testing wool tufts were attached to the tail assembly to determine if the Siemens "X" shaped radar array (Higschgeweih) created any undesirable air flow around the tail assembly. It did not, however, speed was reduced by 50 miles/hour.

A close up of the black colored tail assembly (RLM 22) with its wool tufts on the Me 262 V2 2nd version. Notice that the Balkenkreuz on the fuselage has been almost eliminated.

Left: Me 262 V2 2nd version appears to the right of the photograph. Probably discussing the flying machine's characteristics with the Siemen's search radar is (far left) Oberstleutnant Heinz Bär, Kommander of III Gruppe of Ergänzungsjädgeschwader 2. Center in civilian clothing is Dipl.-Ing. Curt Zeiler of Messerschmitt AG, a specialist in Me 262 flight characteristics. Photo from late 1944. Below: The Me 262 V2 2nd edition appearing to the far right of photo as Heinz Bär (left) and Curt Zeiler (wearing eyeglasses) discuss some features of the Me 262's starboard wing trailing edge controls. Photo from late 1944.

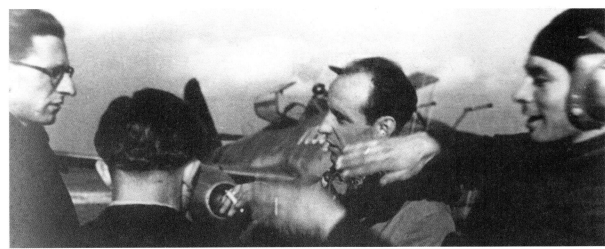

With the Me 262 V2 2nd version in the background, its test pilot, Heinz Herlitzius (right) of Messerschmitt AG, appears to be in highly animated discussion with Curt Zeiler (left) from Messerschmitt AG's design group. Photo from late 1944.

Inside the hatches for the Rheinmetall-Borsig MK 108 30mm machine cannon were located electronics for Siemens' FuG 218 search radar.

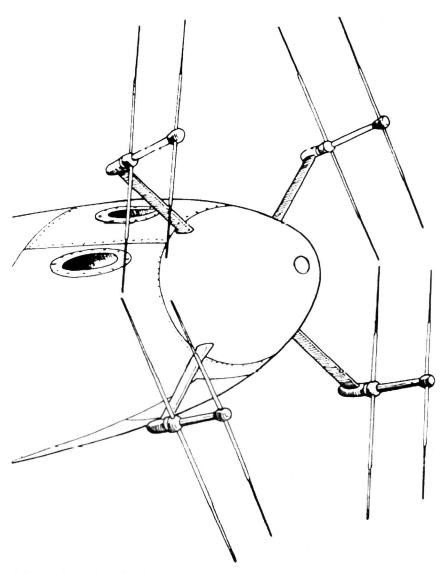

A close up illustration of the Siemens' (Funk Gerät) FuG 218 antenna array mounted on the Me 262 V2 2nd version.

A close up of Me 262 V2 2nd version and its Siemens (Funk Gerät) FuG 218 search radar antenna array as seen from its starboard side nose. The round circle in the nose cone is the lens opening for its gun camera.

All of Me 262 V2 2nd version's testing of the Siemens' FuG 218 search radar array was to perfect it for the serial production of the two man Me 262 B-1a/U1 night fighter. Water color by Piotr Lopalewski.

The Me 262 V2 2nd version seen during its testing of the Siemen's FuG 218 search radar. Both hatches for access to the Rheinmetall-Borsig MK 108 30mm machine cannon are shown wide open.

Me 262 V2-2 [V056]
with a Siemens FuG 218 Search Radar Array and
Port Wing Lorenz FuG 226 "Neuling" IFF Mast Antenna

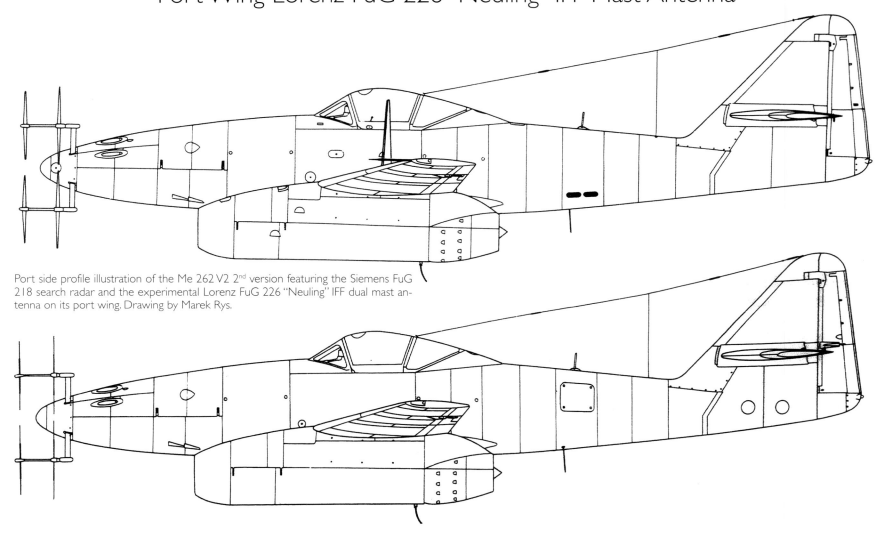

Port side profile illustration of the Me 262 V2 2nd version featuring the Siemens FuG 218 search radar and the experimental Lorenz FuG 226 "Neuling" IFF dual mast antenna on its port wing. Drawing by Marek Rys.

Port side fuselage profile of the proposed single seat Me 262 A-1a night fighter equipped with a Siemens (Funk Gerät) FuG 218 search radar. Drawing by Marek Rys.

A two mast vertical antenna for the Lorenz FuG 226 "Neuling" IFF radar seen on a flying machine's port wing. This device responded by ground control radars.

The Me 262 V2 2nd version. It is seen fitted with its Seimens FuG 218 search radar and notice, also, seen on its port wing are the dual mast antenna of the Lorenz FuG 226 IFF system.

Below: A nose on illustration of the Me 262 V2 2nd version featuring its nose mounted Siemens (Funk Gerät) FuG 218 search radar array and the FuG 226 IFF dual mast antenna mounted on its port upper wing. Illustration by Marek Rys.

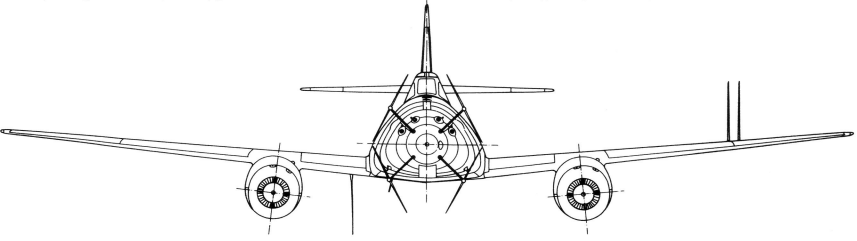

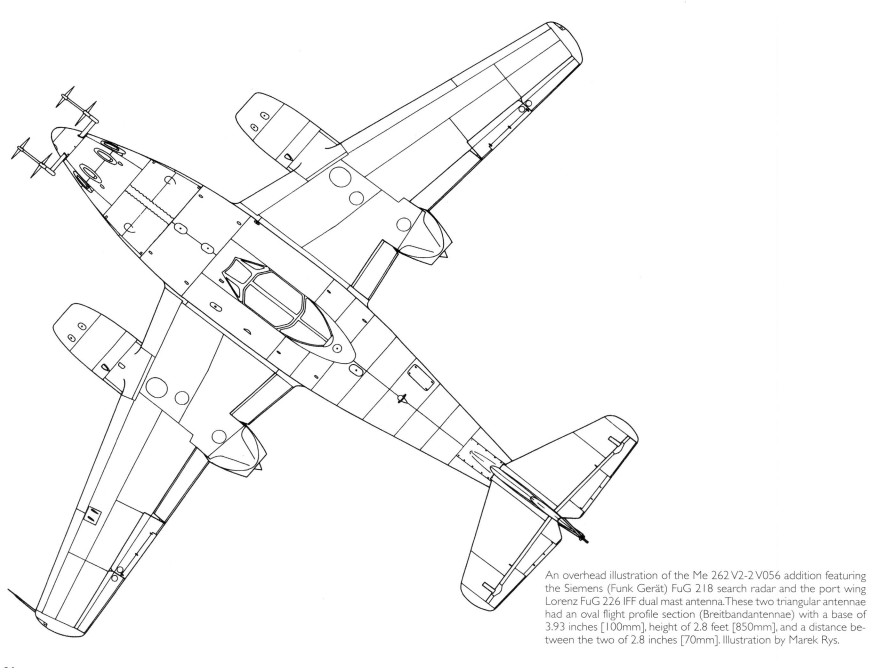

An overhead illustration of the Me 262 V2-2 V056 addition featuring the Siemens (Funk Gerät) FuG 218 search radar and the port wing Lorenz FuG 226 IFF dual mast antenna. These two triangular antennae had an oval flight profile section (Breitbandantennae) with a base of 3.93 inches [100mm], height of 2.8 feet [850mm], and a distance between the two of 2.8 inches [70mm]. Illustration by Marek Rys.

An Me 262 A-1a fighter/bomber. What makes this nose on photo very interesting is what appears to be a dual Lorenz FuG 226 "Neuling" IFF mast antenna out on its port wing.

Me 262 V2-2 [V056]
with a Siemens (Funk Gerät) FuG 218 Nose Antenna Array and
a Lorenz FuG 226 "Neuling" IFF Port Wing Single Mast Antenna

The Me 262 V2 2nd looking pretty much beat up, however, several interesting items appear on this tired and well used V056 airframe. First, notice that it carries a single Lorenz IFF mast antenna on its port side upper wing. Second, a short dorsal fin appears aft the cockpit canopy. This dorsal fin appears larger than other known experiments. This experiment with the higher/shorter dorsal fin is unclear.

Me 262 V2-2 [V056]
with Experimental Dorsal Fin

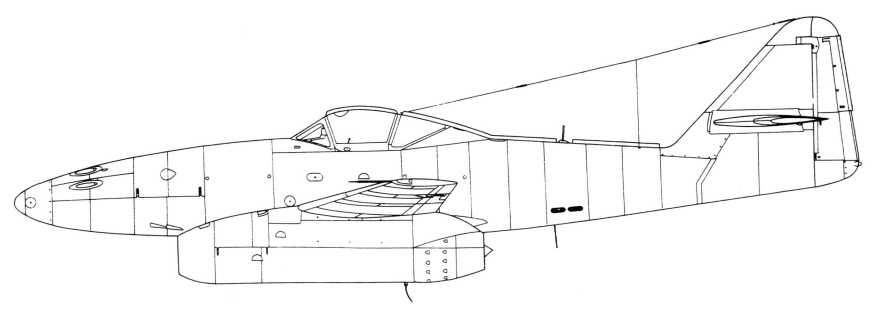

Line drawing of the [V056] by Marek Rys showing it with a dorsal fin running from the top of its cockpit canopy back to the base of the vertical fin.

Based on this photo and its black tail assembly, it appears that [V056]'s next job was to be fitted with a full length short height dorsal fin in an attempt to improve its directional stability.

It appears that Messerschmitt AG engineers tried several experimental dorsal fins on the Me 262 V2 2nd edition [V056]. Here, for example, is a photo of the [V056] pretty badly beating up seen at war's end, however, it features a partial yet much taller dorsal fin than previously thought. This purpose/results of this experiment is unclear.

Me 262 V2-2 [V056]
Initial Experiments with a Lowered Vertical Stabilizer-Rudder-Trim Tab

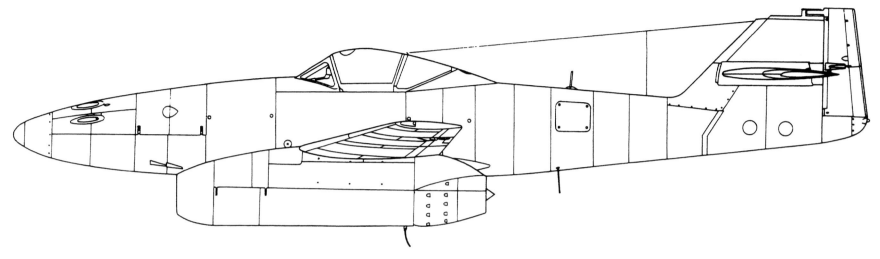

Port side fuselage profile illustration of Me 262 V2-2 [V056] featuring its shortened vertical stabilizer. Drawing by Marek Rys.

The Me 262 V2 2nd version [V056] with and the first attempt at lowering the vertical stabilizer. Messerschmitt AG design engineers were seeking to achieve better directional stability by shortening the vertical stabilizer. Crash landing on the grass December 12th 1944. It is unclear if the shortened vertical stabilizer contributed to this landing accident. Notice that the starboard tire has completely left the rim.

The Me 262 V2 2nd [V056] version's tail assembly before modifications began, although duct tape appears to have been applied over the metal joints.

A view of the Me 262 V2 2nd [V056] version's vertical stabilizer with its leading edge simply removed and duct tape applied. Photo features the flying machine's rear starboard side.

Another view of the Me 262 V2 2nd [V056] version with its shortened vertical stabilizer but this time as seen from its rear port side.

A rear view of Me 262 V2 2nd [V056] version after its landing accident on December 11th 1944. It was repaired and the tail assembly further modified and tested. It is unclear when [V056] was placed on the scrap heap at Lager Lechfeld.

Me 262 V2-2 [V056]
Final Experiments with a Lowered Vertical Stabilizer-Rudder-Trim Tab

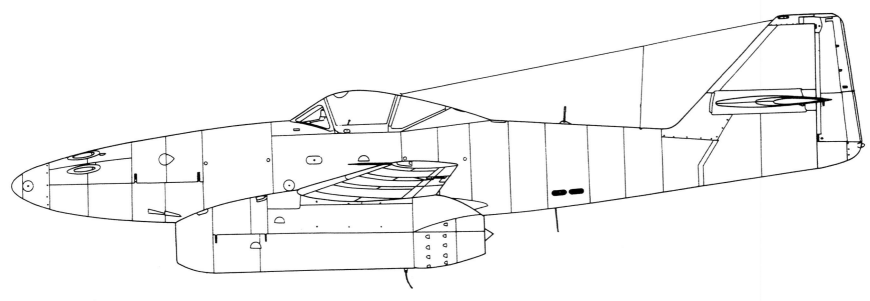

The Me 262 V2 2nd [V056] version as seen from its port side by Marek Rys.

Perhaps one of the last photographs taken of the Me 262 V2 2nd version [V056]'s final modified vertical tail assembly. (This flying machine is seen in the upper right corner of the photograph.) This may have been its last duties of the war because it was found post war in a dump along with several other Me 262 Versuchs flying machines, such as the Me 262 V9 seen in the lower right corner of the photograph.

A close up of the Me 262 V2 2nd version [V056] (right) and its modified tail assembly.

What appears in the final end of Me 262 V2 2nd version [V056] as seen in the upper right corner of the photograph...or is it? Could it be Me 262 V1 Stage 2? If this vertical stabilizer belongs to [V056] then it ended its research duties with the modified vertical stabilizer. At war's end this and several other Me 262s were found at Lager Lechfeld. Me 262 V9 (see p 95) is seen at the lower right corner of the photograph.

An Me 262 A-1a with Werknummer 501228 featuring some of the applied research from the tail assembly modified on the Me 262 V2 2nd version [V056]. This unfinished series production 262 A-1a is seen here with an enlarged trim tab. The flying machine was discovered on May 25th 1945 at Obertraubling.

Me 262 V3

- Me 262 V3...the third prototype.
- Werknummer 262 000 03.
- Radio call code PC+UC.
- No other letters or numerals appear on Me 262 V3's fuselage.
- Surface paint included RLM 74 dark gray, 75 medium gray mottled upper surfaces with pale blue (weissblau) under surfaces.
- Powered by two Junkers Jumo 004A-0 axial flow gas turbine engines. These engines were delivered to Messerschmitt AG-Augsburg on June 1st 1942.
- First flight of Me 262 V3 started on July 18th 1942, at Leipheim Flugpltaz.
- Flugkapitän Fritz Wendel made a total of six test flights before any other test pilot tried. Dipl.-Ing. Heinrich Beauvais was next. He was chief assessor for fighter aircraft at Erprobungsstelle-Rechin.
- On August 11th 1942, Heinrich Beauvais came to Augsburg to test fly the Me 262 V3. However, he had not mastered the technique of momentarily applying the main wheel's brakes in order to lift the tail off the tarmac so that its control surfaces would function in the air stream coming off its forward motion. Beauvais ran out of runway ending up in a cornfield with the Me 262 V3 suffering severe damage. Both of the gas turbines were torn from their mountings. Both main wheels were also torn away. Me 262 V3 was repaired to flight worthy status.
- On September 20th 1943, test pilot Gerd Lindner reportedly reached an unofficial world airspeed record of 596.5 miles/hour [960 kilometers/hour] at 16,404 feet [5,000 meters]. Lindner reported that at 596.5 miles/hour he experienced severe vibrations in the aft fuselage and a general instability.
- Messerschmitt AG designer/engineers believed that the vibrations and general instability experienced by Gerd Lindner were due to disrupting air flow over the cockpit affecting the tail assembly.
- In December 1943, two months after Gerd Lindner's high speed experience with the Me 262 V3, Messerschmitt designers/engineers attached cotton tufts aft the cockpit canopy, attached a movie camera, to determine the nature of air flown over the cockpit.
- After the first flights in December 1943, the Me 262 V3 was modified with a more streamlined cockpit canopy and the flying machine was flight tested again with cotton tufts. However, it appeared that the modified cockpit canopy did not appreciably change the air flow over cockpit then on to the tail assembly for the better.
- Final disposition of the Me 262 V3 is unclear. It was reportedly destroyed during a USAAF bombing raid on September 12th 1944.

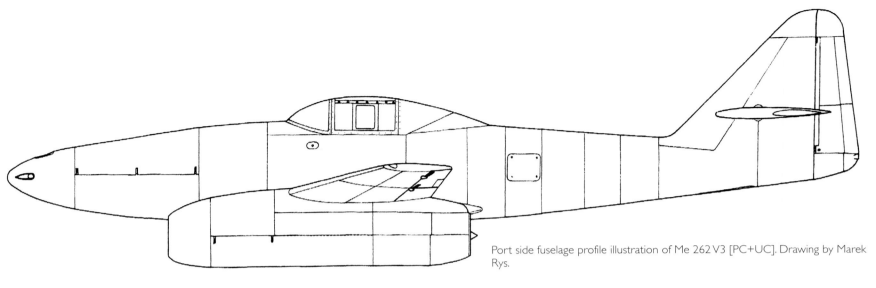

Port side fuselage profile illustration of Me 262 V3 [PC+UC]. Drawing by Marek Rys.

Dipl.-Ing. Heinrich Beauvais (right) and Hauptmann (captain) Otto Brehens, director of fighter testing at the Rechlin Experimental Test Center (left).

Gerd Lindner, on left, piloted the Me 262 V3 to an unofficial air speed of 596.5 miles/hour [960 kilometers/hour] on September 20th 1943.

The third Versuchsflugzeug prototype continued to use the tail wheel favored by Willy Messerschmitt. Drawing by Marek Rys.

An overall view of the Me 262 V3 with radio call code PC+UC and prototype Junkers Jumo T1 (later designated 004A) gas turbines. Scale model and photograph by Günter Sengfelder.

The Me 262 V3 shown parked out on the tarmac with its starboard side Junkers Jumo 004 T1 gas turbine completely covered with a tarp. Messerschmitt AG designers/engineers appear to be concerned about some feature of the starboard wing and its Jumo 004 attachment hardware.

The Me 262 V3 featuring a close up view of its semi swept back wing (straight inner wing and 18 degree swept outer wing) with radio call code PC+UC and prototype Junkers Jumo T1 (later designated 004A) gas turbines. Scale model and photograph by Günter Sengfelder.

A nice view of the Me 262 V3 as seen from its starboard side. Notice the wrap-around windscreen. The cockpit canopy had a one piece glass on its starboard side while the port side canopy contained three square glass panels. Notice the full down position of the landing flap.

A rear starboard side view of the Me 262 V3. In the background is a multi-piston engined Messerschmitt Me 323 "Gigant" transport.

A rear port side view of the Me 262 V3. During flight trials, Flugkapitän Fritz Wendel reached a forward speed of 602.7 miles/hour [970 kilometers/hour] in September 1943, and 621.3 miles/hour [1,000 kilometers/hour] in May 1944.

A full starboard side view of the Me 262 V3. Flugkapitän Fritz Wendel is climbing out of the cockpit. Notice that both Junkers Jumo T1 prototype gas turbines (004A) have a wire screen fitted to their air intakes to minimize large objects being sucked into the compressor.

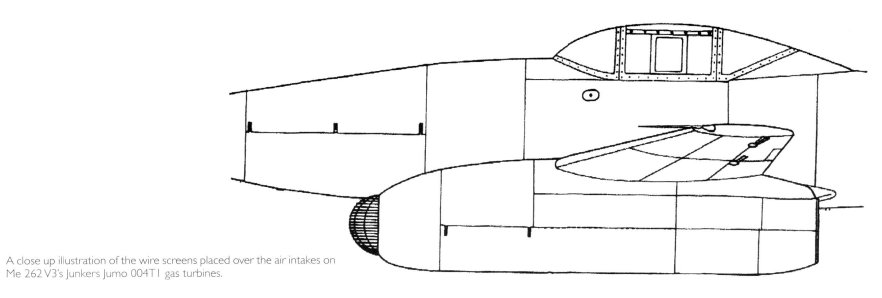

A close up illustration of the wire screens placed over the air intakes on Me 262 V3's Junkers Jumo 004T1 gas turbines.

A close up of Messerschmitt AG test pilot Flugkäptan Fritz Wendel (1915-1975) looking very dapper with his ascot tie and leaning up against the aft starboard fuselage of the Me 262 V3...which he just dismounted. His first test flight in this 3rd prototype [PC+UC] was on July 18th 1942.

Above: A close up of Fritz Wendel seen climbing out of Me 262 V3's cockpit at Lager Lechfeld on March 20th 1943. Notice the clear top panel on the cockpit canopy.

Right: A close up of the Me 262 V3 [PC+UC] as seen from its rear port side and featuring its cockpit canopy with its hinged center panel. Notice, too, that the glass aft the cockpit canopy wraps around in a fashion similar to the wind screen.

A nice overhead view of the Me 262 V3 [PC+UC] featuring the entire length of its fuselage. A mechanic is out on the port side wing appearing to be performing maintenance on the prototype Junkers Jumo T1 gas turbine.

A close up of the Me 262 V3's cockpit and tail assembly as seen from its port side. In the background is a Messerschmitt Me 321 "Gigant" glider transport.

Fritz Wendel bringing the Me 262 V3 [PC+UC] back to Lager Lechfeld after a test flight. Me 262 V3 was lost on September 12th 1944, from a USAAF bombing raid.

The Me 262 V3 [PC+UC] coming in for a landing. Smoke appears billowing from the exhaust end of the gas turbines...from unburnt fuel which collected in the engine cowling and ignited upon landing.

The Me 262 V3 [PC+UC] rolls to a stop with test pilot Flugkapitän Fritz Wendel inside the cockpit, as seen from the Me 262 V3's starboard side in this photo.

One of the "black suited" mechanics leaving the Me 262 V3's cockpit after performing instrument maintenance. Notice how the cockpit canopy is hinged to open over to starboard.

Me 262 V3's damaged aft fuselage, outlined in white chalk, resulting from high speed tests by Gerd Lindner...notice the rivets popping out and general deformation of the aft fuselage aluminum skin.

A close up of Me 262 V3's deformation due to high speed flight testing by Gerd Lindner.

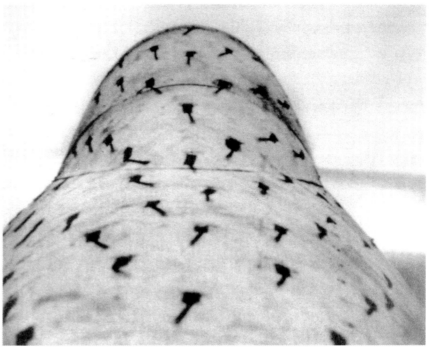

Me 262 V3 [PC+UC] aerodynamic testing to determine air flow over the cockpit canopy. Wool tufts had been attached to the cockpit canopy before any modifications were carried out. The purpose of this testing is to improve air flow around the tail assembly...a continuing effort to improve this problem area right up to war's end.

Modifications made to the Me 262 V3 [PC+UC]'s aft cockpit canopy which included a longer fairing going further aft toward the tail in an attempt to improve air flow over the tail assembly. Wool tufts have been attached to determine what effect, if any, has resulted from the modifications.

A nose port side view of the Me 262 V3 [PC+UC] from the ground featuring its twin Junkers Jumo 004AO gas turbines. Scale model and photograph by Günter Sengfelder.

The Me 262 V3 [PC+UC] featuring a nice view of its landing gear doors. Scale model and photograph by Günter Sengfelder.

A well used Me 262 V3 photographed at Lager Lechfeld from atop a hangar. White stains on the grass may have come from the V3's gas turbines during test runs. In the background is a Messerschmitt Me 321 glider transport.

Me 262 V4

- Werknummer 262 000 004.
- Radio call code PC+UD.
- No other letters or numerals appear on the fuselage.
- Nose cone painted RLM 70 black-green.
- Powered by twin Junkers Jumo 004A-O axial flow gas turbine engines.
- First flight occurred on May 15th 1943.

- This airframe was used to experiment with the Me 262 A-1a's ability to carry external ordnance (bombs).
- Suffered a crash landing on its 2nd test flight at Schkeuditz on July 23rd 1943 or July 26th 1943 with test pilot Gerd Lindner at the controls when. Me 262 V4's nose wheel collapsed. It suffered an estimated 60% damage and was written off as unrepairable.
- Prior to its crash landing, Me 262 V4 [PC+UD] had completed 51 flight tests.

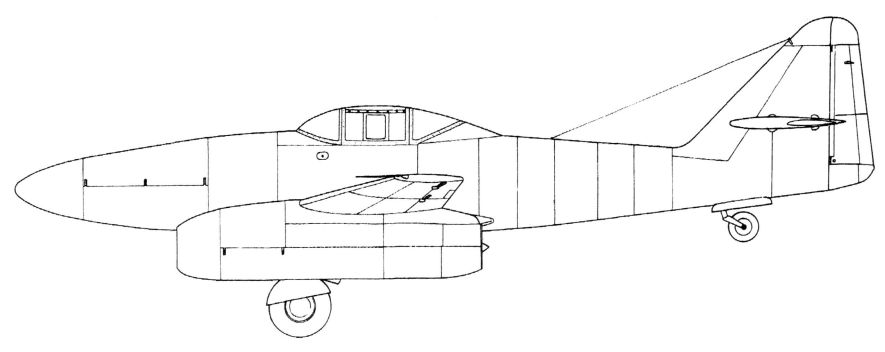

Port side fuselage profile illustration of the Me 262 V4 [PC+UD]. Drawing by Marek Rys.

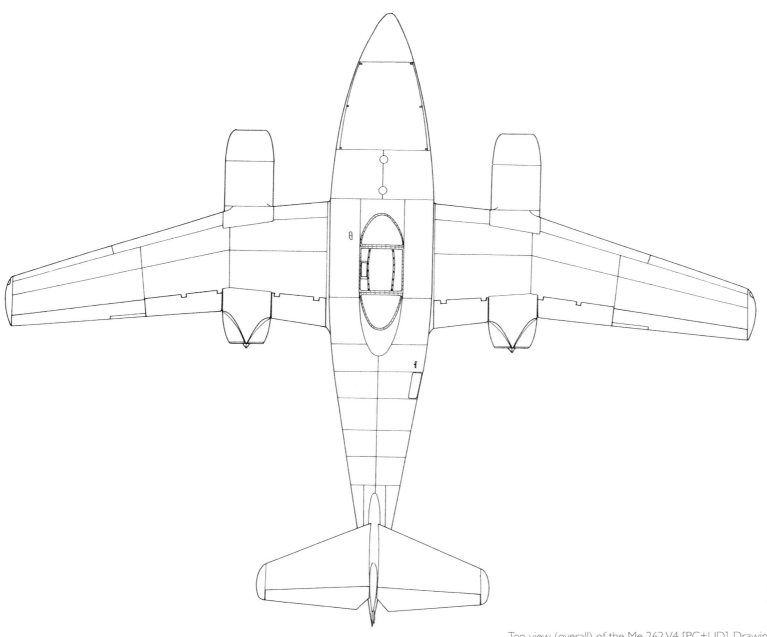

Top view (overall) of the Me 262 V4 [PC+UD]. Drawing by Marek Rys.

Me 262 V5

- Werknummer 262 000 05.
- Although this research prototype is considered number 5, it was similar to Me 262 prototypes V3 and V4, except that V5, like V4, was also powered by twin Junkers Jumo 004A-O axial flow gas turbine engines. Me 262 V3 still used the Jumo 004A-O gas turbines.
- Construction of this prototype reportedly took two years to complete.
- The Me 262 V5 [PC+UE] was the first Me 262 to have a tricycle landing gear, although this experimental gear was fixed during its test flights because its hydraulic retraction system was not yet functional.
- Nose wheel came from the Me 309.

- Wing leading edge slots installed and flight tested.
- Nose wheel damaged during a hard land ing/accident on August 4th 1943. Certified flight worthy in January 1944.
- Served as a test aircraft with two experimental Rheinmetall-Borsig R502 take off assist rockets with their 1,102 pound [500 kilogram] thrust for 6 seconds.
- Experienced a second landing accident due to a blown tire on its first flight after being repaired ... January 2nd 1944. Some reports state that its second landing accident occurred on February 1st 1944. Nevertheless, it appears that the Me 262 V5 was never repaired again.

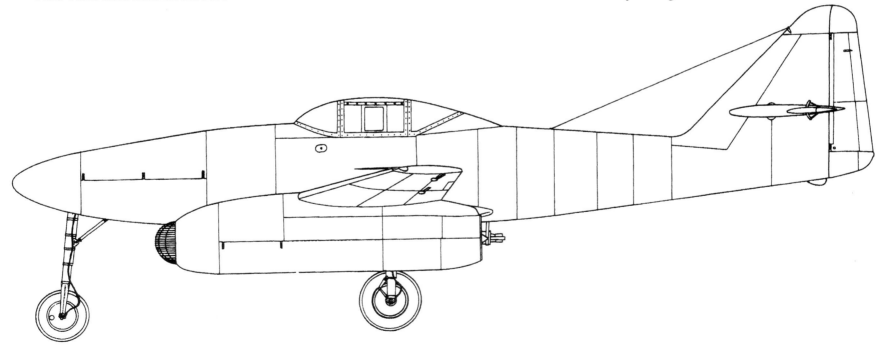

Port side fuselage profile illustration of the Me 262 V5 [PC+UE]. Drawing by Marek Rys.

The Me 262 V5 carrying radio call code PC+UE. Werknummer 262000 05 as seen from its rear starboard side. This flying machine was the first Me 262 with a nose wheel although all the wheels did not yet retract. Notice that it has not yet been fitted with the Rheinmetall-Borsig R-502 take off assist solid fuel rockets.

The Me 262 V5 with radio call code PC+UE as seen from its nose starboard side. The V5 was a test bed for the tricycle landing gear. This machine also had two Rheinmetall-Borsig R-109-502 solid rocket assisted take off units.

This Me 309 V1 in July 1942 was Messerschmitt AG's initial attempt at a nose wheel equipped flying machine. The Me 262 V5, the first Me 262 with a nose wheel, did not make its first test flight until June 6th 1943!

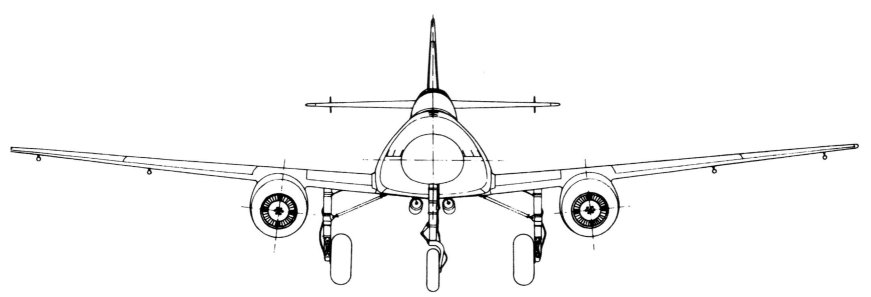

An early drawing from November 26th 1942, featuring a Me 262 with nose wheel. Notice the twin Rheinmetall-Borsig R-502 take off assist rocket units under the fuselage. Courtesy: Me 262 Volume One, J. Richard Smith and Eddie J. Creek, Classic Publications, 1997.

Two Rheinmetall-Borsig R-502 solid fuel take off assist rockets of the type attached to the Me 262 V5. The wire hanging down is the electrical connection.

The Me 262 V5 with its fixed tricycle undercarriage and a good view of its twin Rheinmetall-Borsig R-502 1,700 pound thrust take off assist rockets.

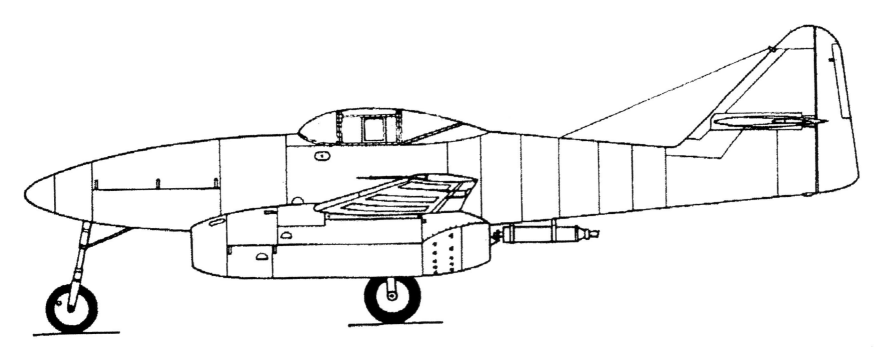

Illustration of the Me 262 V5 with its Rheinmetall-Borsig R-502 take off assist rockets by Marek Rys.

The Rheinmetall-Borsig R-502 solid fuel take off assist rocket as installed on the Me 262 V5. It weighed 112 pounds [51.5 kilograms] and produced a thrust of 1,700 pounds [771 kilograms] for 7.5 seconds.

The Me 262 V5 has started its take off roll. Notice the pure while flame exiting from its twin Rheinmetall-Borsig take off assist rockets.

The Me 262 V5 making its first flight on June 6th 1943. It had a fixed tricycle undercarriage and two Rheinmetall-Borsig R-502 take off assist rockets to help it lift off sooner than the twin Junkers Jumo 004A-O prototype gas turbines could then provide. Actually, one of the main disadvantages right up through war's end of the Me 262 A-1a was that it required a very long take off run before lift off.

Me 262 V5-2 [V167]

- The second Me 262V5 labeled "Versuchs" and known by the last three digits of its Werknummer, or "V167," painted in white. This occurred after the loss of Me 262 V5 Werknummer 000 0005;
- Werknummer 130167;
- Radio call code SQ+WF but not normally seen on the flying machine;
- First flight on May 31ˢᵗ 1944;
- Reportedly flew 303 test flights up through March 1945, for a total of 50 hours and 25 minutes;
- V167 was used to test undercarriage modifications, cabin heating and a window washing mechanism;
- V167 was also used to flight test external bombs carried on twin Messerschmitt AG designed Wikingerschiff and Focke-Wulf designed ETC 503 bomb racks;

- Used to test the "Adler" or eagle EZ 42 or Eilnheitszielvorrichtung automatic gyroscopic gun sighting device;
- V167 is known to have been used to test take off assist rockets;
- It is the only known Me 262 to have a redesigned rudder without a counter balance horn. This proved disappointing in flight tests;
- Generally photographed in a camouflage paint scheme of RLM pale blue 76 undersides, RLM dark greens 81 and 82 overall;
- Fitted with a BSK 16 airborne gun camera;
- V167 is known to have been demonstrated to a Japanese aviation delegation;
- Final disposition at war's end is unclear.

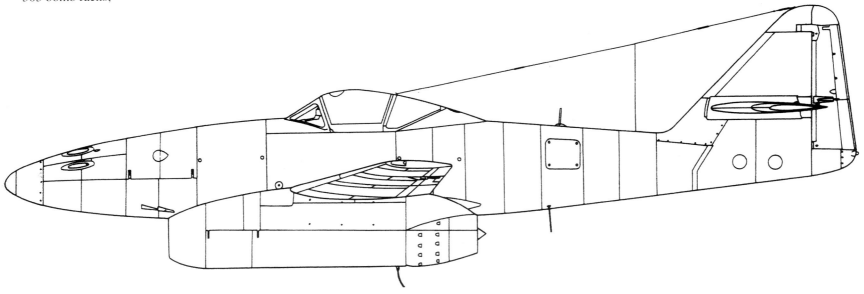

Port side profile illustration of the Me 262 V5-2. Drawing by Marek Rys.

A nose on view of the Me 262 V5-2 [V167] as delivered fresh from Messerschmitt AG. It was initially painted overall in RLM 76 "weissblau" with radio call code SQ+WF. It looked and appeared very much like its sister "S" types. Notice, too, the lack of a BSK-16 gun camera port in the nose cone. First flight of SQ+WF was on May 31st 1944.

The Me 262 V5-2, Werknummer 130167, SQ+WF in its coat of RLM 76 Weissblau camouflage. This flying machine was assembled at Messerschmitt AG's Schwäbisch Hall facility.

Messerschmitt AG test pilot Dipl.-Ing. Flugkapitän Karl Baur (1913-1963) seen in the cockpit of Me 262 V5-2. He is testing the new gun sight known as the electronic EZ 42 "Alder," or eagle sighting system. This machine cannon/gun sight was a disappointment. Notice that the former SQ+WF has been given a new paint job...RLM 76, 81, 83.

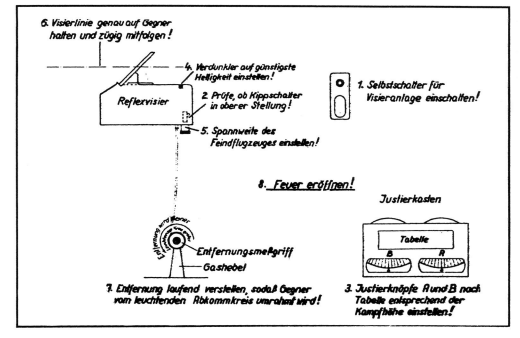

A schematic drawing featuring the Askania EZ 42 electronic sighting system tested in the Me 262 V5-2. Its goal was to allow the pilot to shoot at an airborne target without allowance for the movement from the fixed machine cannon in the Me 262's nose.

121

The newly repainted Me 262 V5-2 as seen from its nose starboard side. It is now camouflaged in RLM 76, 81, and 83. Its only markings, in addition to the Balkenkreuz of the B3 style, is the white "V167" painted on the forward fuselage.

A movie still photo of the Me 262 V5-2 [V167] Werknummer 130167 from December 1944, while it was participating in fighter/bomber trials.

Above Left: It is claimed that Me 262 V5-2 [V167] made 303 test flights. If so, then this flying machine was flown more times than any other Me 262...test versions or fighters. The [V167] was also demonstrated to a Japanese aviation delegation. Above: Me 262 V5-2 [V167] seen lifting off for one of its numerous test flights with Gerd Lindner (?-1945) at the controls...perhaps the most experienced of the Messerschmitt AG Me 262 test pilot at war's end. Lindner died post war in 1945 in an automobile accident. Above Right: The Me 262 V5-2 [V167]'s tail assembly featured here shows no counter balance horn on its elevators. The test was disappointing and removing the counter balance was never adopted for production Me 262 A-1as.

A nice nose port side ground level view of the Me 262 V5-2 [V167]. It appears that the pilot is being assisted by a member of the ground crew prior to a test flight.

The Me 262 V5-2 [V167] seen prior to take off. Notice the absence of any bomb racks although cannon ports for its four Rheinmetall-Borsig MK 108 30 mm machine cannon are evident.

Me 262 V5-2 [V167] featuring its nearly 4 foot [1200mm] long nose probe. Mounted at the end was a "yaw angle measuring" device. The probe was mounted prior to its chopped/shortened vertical stabilizer.

Mentioned earlier, the Me 262 V5-2 reportedly made 300 plus test flights...mainly for stability investigations, external bomb loads, and in early 1945, as a test machine for the newly designed nose wheel fork.

The Me 262 V5-2 [V167] with its four foot [1200mm] long nose probe. Left to right is Gerd Lindner, Willy Messerschmitt, Werner Thierfelder, and Gerhard Caroli. Thierfelder was the commander of EKdo 262.

Messerschmitt AG's most experienced Me 262 test pilot, Gerd Lindner has his hand on the nose probe of the Me 262 V5-2 [V167]. In the background with a collapsed nose wheel is Me 262 S-6. The block and tackle seen above V5-2's nose probe is to be used to lift its nose off the ground. Willy Messerschmitt appears far left in the photo. The Luftwaffe officer is Hauptmann (Captain) Werner Thierfelder, commander of EKdo 262, who would later lose his life in the S-6 on July 18th 1944.

Me 262 V6

- Werknummer 130001.
- Radio call code VI+AA.
- Powered by twin Junkers Jumo 004B-1 axial flow gas turbine engines. The Me 262 V6 had improved aerodynamic cowlings around its engines.
- Painted overall in RLM 76 weissblau or pale blue.
- Its "V6" located on both sides of the vertical fin was painted black.
- VI+AA's first flight occurred on October 17th 1943. Its test pilot was Feldwebel (Technical Sargent) Gerd Lindner.
- The Me 262 V6 was the first Me 262 to have a fully functional hydraulic retracting tricycle landing gear. It would be a standard feature for all Me 262s to follow.

- Flight testing involved extending its landing gear at various forward speeds while airborne such as 342 miles/hour [550 kilometers/hour] for purposes of determining if gear extension would be useful as an air brake. Test pilots found that gear could be lowered at speed with only minimal changes in trim.
- Overall, the Me 262 V6's outward appearance did not vary much from the upcoming Me 262 A-1a series, except for its old style cockpit canopy and tail skid.
- The Me 262 V6 was the first prototype to have the ability to have four Rhein-metall-Borsig MK 108 30mm machine cannon installed in its nose, although cannon were never installed in the Me 262 V6.
- The Me 262 V6 was destroyed in a crash landing on March 9th 1944, killing its test pilot, Feldwebel Kurt Schmidt.

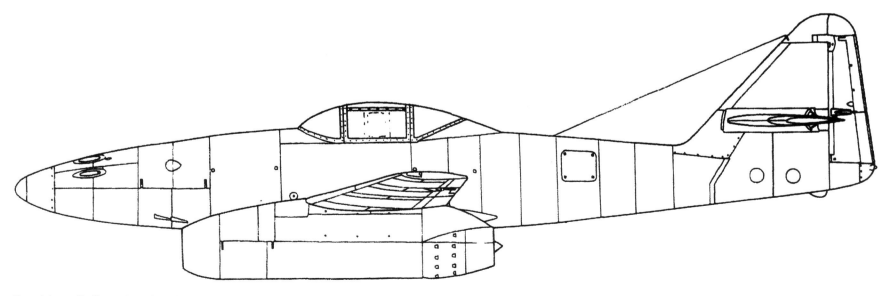

Port side profile illustration of Me 262 V6 [VI+AA]. Drawing by Marek Rys.

The Me 262 V6 with radio call code VI+AA. It was the first Me 262 to be fitted with a fully retractable tricycle undercarriage.

The Me 262 V6. Notice its protective skid under the rudder on the tail assembly...to protect the aft fuselage from damage should it scrape its tail on the runway during take off and/or landing.

The Me 262 V6 was painted in RLM 76 weissblau with its "V6" painted black.

Me 262 V6, with its twin Junkers Jumo 004 B-1 axial flow gas turbines, made its first flight on December 20th 1943, with Oberst (Colonel) Dietrich Peltz at the controls. It was lost on March 9th 1944 when it crashed on landing, killing its pilot Feldwebel (Technical Sargent) Kurt Schmidt.

A rear starboard side view of the Me 262 V6 [VI+AA] taxiing to its take off position on the tarmac at Lager Lechfeld air station for a demonstration for Reichsmarschall Hermann Göring and his entourage from the RLM.

A demonstration of Willy Messerschmitt's Me 262 for Luftwaffe leader Hermann Göring...featuring the Me 262 V6, which has just landed. This demonstration took place on November 2nd 1943 at Lager Lechfeld. Reichsmarschall Hermann Göring, dressed in his white uniform, is passing by the V6's nose, while Willy Messerschmitt appears to be inspecting the air intake of the port Junkers Jumo 004 gas turbine.

The Me 262 V6's nose wheel seen from its port side. This nose wheel had an external shimmy brake and scissors and a steering arm. The external shimmy brake was eliminated on the Me 262 V7...and was placed internally.

A close up of the Me 262 V6's nose wheel as viewed from its port side. All Me 262 A-1as were equipped with a shimmy brake and a steering limiter. Photo from April 1944.

A close up of the Me 262 V6's shimmy brake with scissors/steering arm forming the linkage between the swivelling wheel fork and ridged cylinder. The Me 262 V6's nose wheel was also equipped with an adjustable locking ring and an internal friction disk.

Me 262 V7

- Werknummer 130002.
- Radio call code VI+AB.
- First flight occurred on December 20th 1943.
- Powered by twin Junkers Jumo 004B-1 axial flow gas turbine engines.
- Me 262 V7 was painted overall in RLM 76 "weissblau" or pale blue. 'W" was applied in white on both sides of the vertical fin near its leading edge.
- The Me 262 V7 was a flying test bed used to improve the belt-fed four Rheinmetall-Borsig MK 108 30mm machine cannon. Extensive flight testing was required because when the Me 262 went into a tight banking turn centrifugal forces tore the MK 108's ammunition feed belt. The belt was later modified, and the Me 262 could make tight turns and still fire its four Rheinmetall-Borsig MK 108 30mm machine cannon without difficulty.
- The Me 262 V7 is similar in many ways to its sister Me 262 V6, with the exception that the V7 had counter balance horns atop its elevators similar to those found on the Me 262 V5.

- Me 262 V7 was damaged in a landing accident on February 21st 1944, after its 17th test flight. This crash landing occurred in a snow storm, plus the Me 262 V7 was having engine trouble. Test pilot Feldwebel (Technical Sargent) Kurt Schmidt belly landed the prototype in the snow. It was repaired and flown again on April 11th 1944.
- During its rebuild the Me 262 V7's cockpit canopy was modified with a new all around view cockpit canopy. This cockpit canopy would later become the standard cockpit for the Me 262 A-1a.
- In addition to a new style cockpit canopy, Me 262 V7's tail assembly was modified.
- During its rebuild, cockpit cabin pressurization was added to Me 262 V7.
- After Me 262 V7's rebuild it was test flown on a regular basis, however, it was destroyed in a crash landing about one month later on May 19th 1944. This crash landing occurred on its 31st test flight, killing its pilot Unteroffizier (Corporal) Kurt Flacks.
- Overall, Me 262 V7 logged a total of 13 hours and 30 minutes flight time.

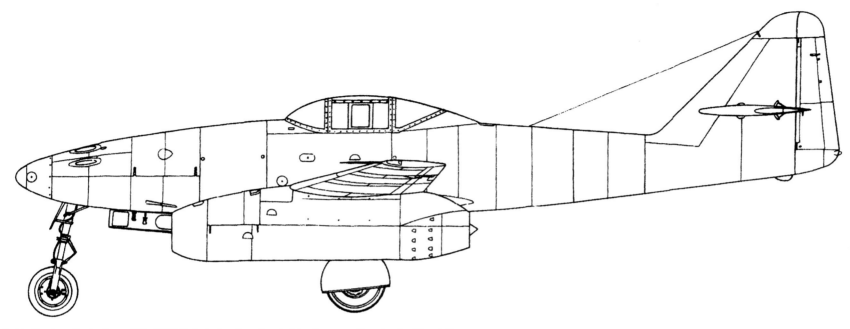

Port side fuselage illustration of Me 262 V6 featuring its extended landing gear. Drawing by Marek Rys.

The Me 262 V7 [VI+AB] Werknummer 130002 photographed during a snow storm and fully covered with snow as it plowed along the runway when it made a wheels up landing on February 21ˢᵗ 1944, due to a reported engine malfunctioning. This accident occurred on its 17ᵗʰ flight. Individuals from Messerschmitt AG surround their Me 262 V7 planning how to recover the flying machine. Notice that both air intakes to the Junkers Jumo 004 gas turbines are completely impacted with snow.

The Me 262 V7 as photographed from its starboard side. It's three panel cockpit canopy lies on at the starboard wing root. The Me 262 V7 would later be rebuilt, however, it would then have the new style bubble cockpit canopy.

The Me 262 V7 as photographed from its port side. Notice that its port wing leading edge slot is in its full extended position.

Opposite: The Me 262 V7 [VI+AA] takes to the air again at Lager Lechfeld after its rebuild. During the rebuild it was equipped with a bubble cockpit canopy and an improved tail assembly. Spring of 1944.

Notice how the snow and dirt have fully impacted the Me 262 V7's port Junkers Jumo 004 gas turbine's air intake.

The end of Me 262 V7. All that remains from the V7's fatal nose dive into the ground on May 19th 1944, at Lager Lechfeld. It was unclear why the Me 262 V7 crashed down on its 31st flight after having logged 13 hours and 30 minutes of flight time.

Me 262 V7-2

- The second Me 262 V7 which was simply designated Me 262 V7-2.
- Werknummer 170303.
- Powered by two Junkers Jumo 004B-1 axial flow gas turbine engines.
- First flight of Me 262 V7-2 was on September 22nd 1944.
- Used extensively to test the external "Wikingerschiff" bomb racks and bomb release mechanisms.
- Used to test the 2,204 pound [1,000 kilogram] thrust solid fuel take off assist Rheinmetall-Borsig R-502 rockets.
- Used to test flight characteristics while operating with only one gas turbine operating.
- Used to test experimental brakes on main wheels.
- Used to test improved cockpit cabin ventilation and visibility.
- Flight tested with a modified tail assembly with control surfaces made out of wood.

- Reportedly made a total of 67 flight tests between September 1944 and February 1945 ... mostly by test pilot Heinz Herlitzius.
- Overall Me 262 V7-2 was flown for a total of 18 hours and 38 minutes.
- Me 262 V7-2, on its 68th flight test, near the end of February 1945, one of its two Rheinmetall-Borsig R-502 take off assist rockets broke loose from its fuselage attachment resulting in considerable fire damage to the tail assembly.
- It is reported that the tail assembly from Me 262 A-1b, the twin BMW 003 powered flying machine (Werknummer 170078) was installed on the Me 262 V7-2. It is unclear when this switch was made, where it was done, and if it was completed prior to war's end. It appears that Me 262 V7-2 never flew again after its 68th test flight.
- At war's end Me 262 V7-2 [V303] ended up in a scrap heap at Lager Lechfeld minus its entire nose assembly.

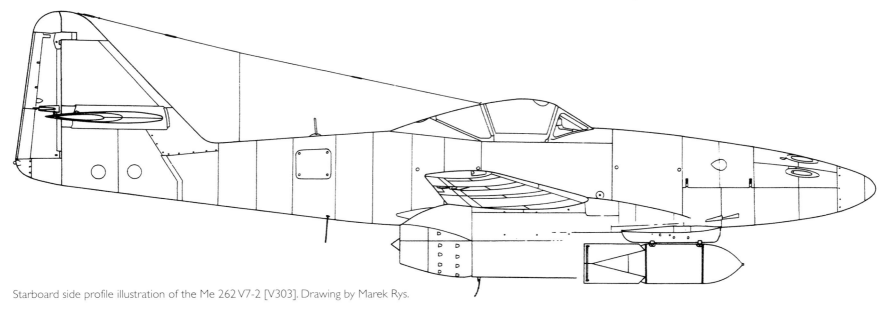

Starboard side profile illustration of the Me 262 V7-2 [V303]. Drawing by Marek Rys.

The Me 262V7-2's four nose Rheinmetall-Borsig MK 108 30mm machine cannon seen fitted with experimental flash suppressers sticking out beyond the fuselage.

Feldwebel (Technical Sargent) Heinz Herlitzius. He piloted the Me 262 V7-2 during most of its 68 test flights.

A nose on view of the Me 262 V7-2 V303 fitted with two Wikingerschiff bomb racks each carrying a 1,212 pound [550 kilogram] bomb. Notice how its four Rheinmetall-Borsig MK 108 30mm machine cannon extend out beyond the nose with its flame suppressers attached.

The Me 262 V7-2 as seen from its nose starboard side. Notice the starboard side "Wikingerschiff" bomb rack and behind are two Rheinmetall-Borsig R-502 solid fuel take off assist rockets.

This photo shows an excellent view of the starboard side "Wikingerschiff" bomb rack installed on the Me 262 V7-2. To the right of the bomb rack is the Me 262's rectangular nose wheel cover door.

A nose on ground level view of the Me 262 V7-2 [V303] and providing a good view of its two 2,204 pound [1,000 kilogram] thrust Rheinmetall-Borsig R-502 take off assist rockets.

One 1,121 pound [550 kilogram] bomb resting on its wooden skid. It is intended for the Me 262 V7-2 [V303] in the background which is being serviced by two black dressed mechanics.

Lager Lechfeld post war. The remains of several important test flying machines. In the foreground is the remains of Me 262 V7-2 [V303]. It is minus its entire nose assembly. Behind it is the Me 262 V9.

Armorers are seen preparing a 1,212 pound [550 kilogram] bomb to be attached to the Me 262 V7-2 waiting in the background. The Me 262 V7-2 has two "Wikingerschiff" bomb racks.

Digital image by Jozef Gatial.

Me 262 V8

- Werknummer 130003;
- Radio call code VI+AC with its prototype number (V8) painted on its nose is large white letters, the only prototype to have had its research number painted on its nose;
- Painted full camouflage except that its "V8" appeared in white characters on both sides of the vertical tail fin near its leading edge;
- Powered by two Junkers Jumo 004 B-1 axial flow gas turbine engines;
- First flight on March 18th 1944;
- Fitted with new cockpit canopy provided overall better view;
- Tested with various armament arrangements in the nose such as the Rheinmetall-Borsig MK 108 30mm machine cannon and the BSK 16 gun camera;
- The first Me 262 to be fitted with four Rheinmetall-Borsig MK 108 30 mm machine cannon and was used for cannon firing trials ... the standard armament for all production Me 262 A-1as to follow;
- Damaged several times during its 258 test flights. Repaired each time;
- April 8th 1944, assigned to Erprobungskommando 262 or Proving Detachment 262. It would later be redesignated Kommando Nowotny;
- Erprobungskommando 262 (EKdo 262) was the formation of a service trials unit located at Lechfeld in Bavaria and commanded by Hauptmann Werner Thierfelder;
- Thierfelder and several others had transferred to Erprobungskommando 262 from 3rd Gruppe of Zerstoeregeschwader 26 which flew the bomber-destroyer version of the Messerschmitt Bf 110;
- Zerstoeregeschwader 26 found the Me 262 A-1 a was an easy machine to fly once they had mastered the problem of throttle handling. With the early gas turbines the throttles had to be advanced very slowly or they were liable to overheat and catch fire.;
- On the other hand, once the pilot had throttled back his gas turbines in preparation for landing at low altitude, he was committed to a landing. If he advanced his throttles and tried to go around again the gas turbines took so long to build up power that the Me 262 A-1a was likely to hit the ground first;
- Pilots who had flown twin engine aircraft such as the Messerschmitt Bf 110 and who had been trained in instrument flying did a better job piloting the Me 262 A-1a with its high speed and short endurance;
- The Me 262 A-1a had sufficient fuel for only 40 to 60 minutes flying time. They conserved fuel by being towed to the take-off line before each flight because they could not afford to spend 10 minutes on the ground wasting precious fuel;
- Believed to have been used by I/KG 51;
- Written off in October 1944 due to a landing accident.

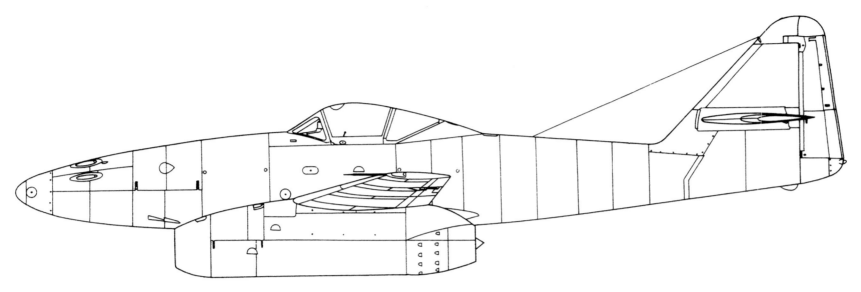

A port side profile illustration of the Me 262 V8 [VI+AC]. Drawing by Marek Rys.

The only known photo of Me 262 V8 (VI+AC). It was used by Erprobungskommando Thierfelder—the fighter test detachment. It also used twin Rheinmetall-Borsig R-502 take off assist rockets.

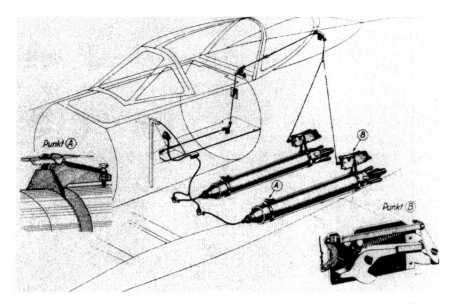

The typical installation arrangement of the Rheinmetall-Borsig R-502 solid fuel take off assistance rocket assembly on the Me 262, such as the Me 262 V8, as seen from its nose port side. A - forward R-502 fuselage attachment hardware connecting to the Me 262's fuselage. B - aft R-502 fuselage attachment hardware connecting to the Me 262s fuselage.

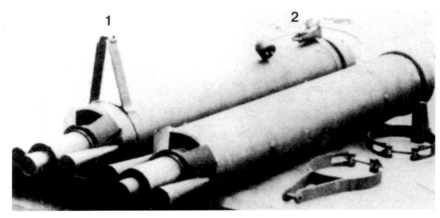

Above: Rheinmetall-Borsig R-502 solid fuel take off assist rockets. To the left are the thrust nozzles. 1) aft attachment hardware; 2) forward attachment hardware. Right: The control panel containing electric switches for the Rheinmetall- Borsig R-502 as seen in the port side cockpit panel of the Me 262 V8.

Me 262 V8-2 [V484]

- The first Me 262 V8 [Werknummer 130003] in October 1944, Werknummer 110484, became Me 262 V8-2, or Versuchs V484;
- It is also known as Me 262 A-1a/U2;
- Purpose of the Me 262 V8-2 was to experiment with an alternative form of a blitzbomber or fast bomber using the TSA 2D bomb aiming device and a man (bombardier) lying prone in a highly modified glass nose of the Me 262 A-1a;
- The Me 262 V8-2 was unarmed the four Rheinmetall-Borsig MK 108 30mm cannon removed when the flying machine was modified to;
- It was believed that the bombing effectiveness of the Me 262 would be far better than the Me 262 A-1a Jabo and Me 262 A-2a blitzbomber versions;
- TSA means "Tief und Sturzfluganlage," meaning low and diving bomb sight;
- Me 262 V8-2's camouflage included upper surfaces in RLM 81 dark green;

- The RLM authorized the modification of a Me 262 A-1a (V484) as the bombardier prototype;
- First flight of Me 262 V8-2 9 (V484)occurred in September 1944. On October 22nd 1944, it had been transferred to Lager Lechfeld and by the end of December 1944, it had completed 22 test flights for a total of 8 hours and 14 minutes;
- Bomb trials included two ETC 504 bomb racks holding [250 kilogram] bombs and dropped at a forward speed of 370 miles/hour [600 kilometers/hour] from 6,600 feet [2,000 meters] altitude;
- Pilot during the trials was flugkapitän Karl Baur and bombardier Bayer;
- A second bombardier carrying prototype Me 262 V11 (V555) was completed but no serial production was ever started;
- Me 262 V8-2 (V484) was ferried to Erprobungsstelle (E-Stelle) Rechlin on January 7th 1945, to undergo further evaluation;
- Final disposition of the Me 262 V8-2 (V484) at war's end is unclear.

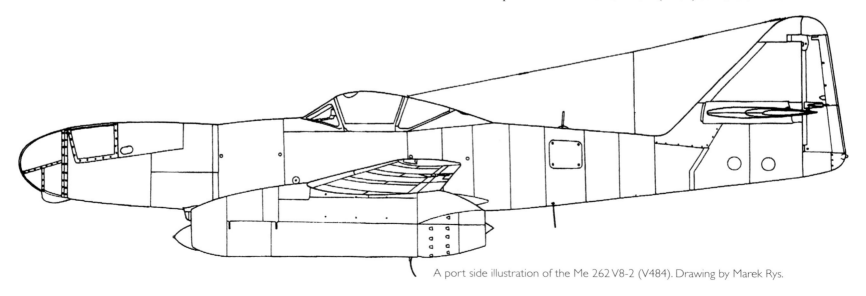

A port side illustration of the Me 262 V8-2 (V484). Drawing by Marek Rys.

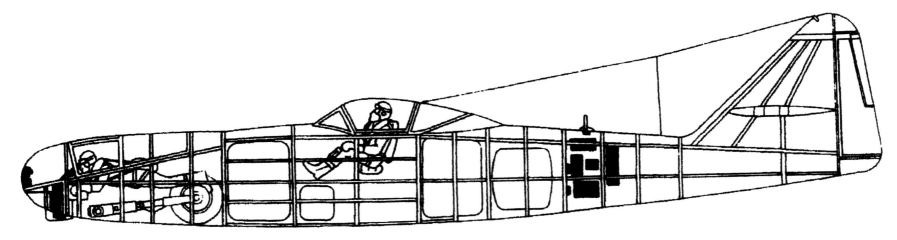

A see through port side illustration of the Me 262 V8-2 (V484) featuring the prone position of the bombardier.

A nose port side view of the Me 262 V8-2 (V484) showing its wooden nose with a glazed Lofte Kanzel I bomb aiming position for the prone laying bombardier.

The Me 262 V8-2 (V484) heavily camouflaged being towed by its nose wheel by a fuel tanker truck. This means of towing had been banned due to the high potential for damaging the nose wheel. Behind the (V484) is the second Me 262 V1 (Werknummer 130015).

A close up of the Me 262 V8-2 (V484) with its glazed nose partially uncovered. Photo reportedly taken at Rechlin Test Center in early 1945, where the prototype blitzbomber was transferred in January 1945.

The box containing the Lofte 7H's protruding lens for its bombsight; the whole thing was attached to the glazed nose of the Me 262 V8-2 (V484). The Lofte 7H is seen from the port side of the Me 262 V8-2 (V484).

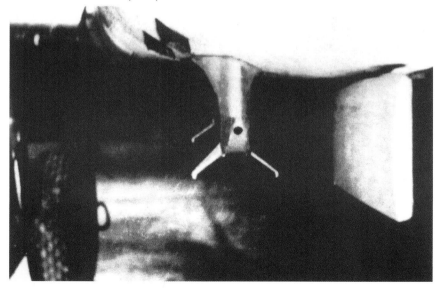

Above: The Me 262 V8-2 (V484)'s glazed wooden frame bombardier's nose as seen from the starboard side. Right: A ETC 504 bomb rack fitted to the Me 262 V8-2 (V484).

Me 262 V9

- Werknummer 130004.
- Radio call code VI+AD.
- Powered by twin Junkers Jumo 004B axial flow gas turbine engines.
- Initially used for flight testing modifications made to perfect its four Rheinmetall-Borsig MK 108 30mm machine cannon. Also used to flight test various radio combinations. Me 262 V9 [VI+AD] would also be used to flight test the proposed Me 262 HG-1 or "home land defender." Flight testing of this Me 262 version reportedly began in October 1944 ... perhaps in January 1945, with a racing cockpit canopy known as "Rennkabine." VI+AD was also flight tested with enlarged vertical and horizontal stabilizers. The horizontal stabilizer had a greater degree of sweep back. Instead of improving the Me 262's directional stability, it made the flying machine even more unstable.
- All the modifications made to Me 262 V9 [VI+AD] were flight tested by Messerschmitt AG factory test pilot Karl Baur. One thing for sure was that he did not like the "Rennkabine" because he kept bumping his head against the glass cockpit canopy.
- By and large, all the modifications made to Me 262 V9 [VI+AD]'s tail assembly were unsatisfactory. Test pilot Baur reported that the modifications caused instability around its yaw axis. The tail assembly was later changed back to its original configuration

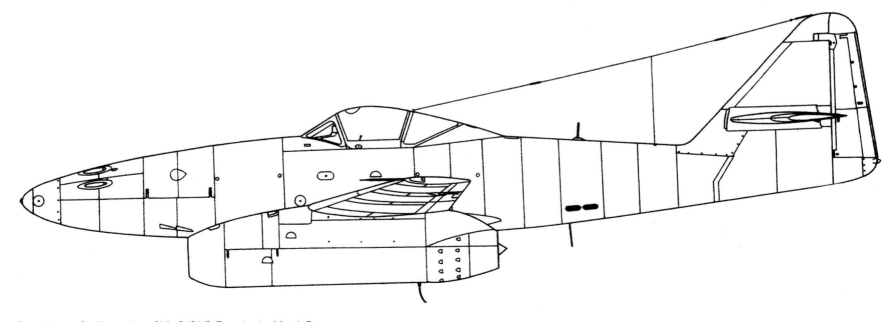

Port side profile illustration of Me 262 V9. Drawing by Marek Rys.

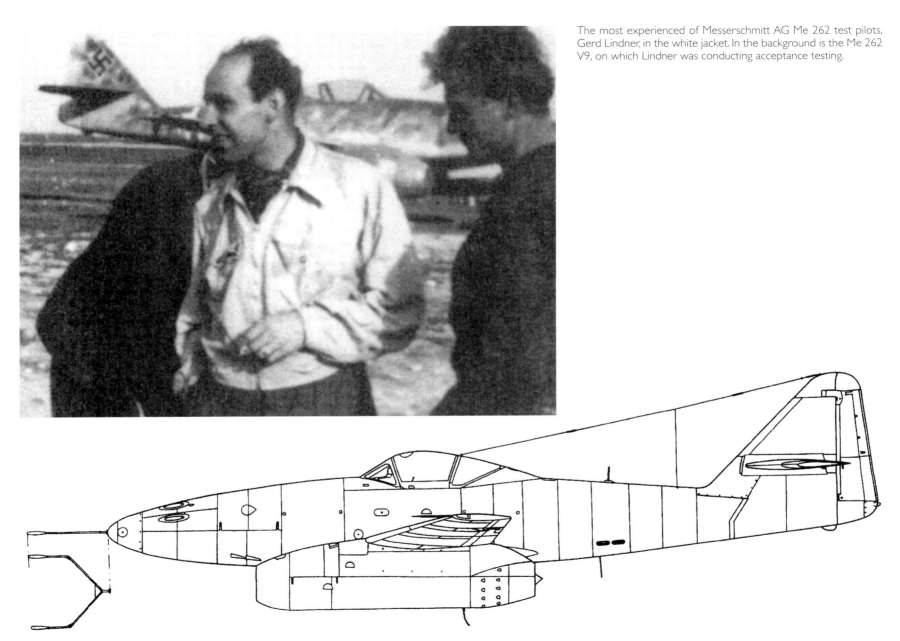

The most experienced of Messerschmitt AG Me 262 test pilots, Gerd Lindner, in the white jacket. In the background is the Me 262 V9, on which Lindner was conducting acceptance testing.

A port side profile illustration of the Me 262 V9 [VI+AD] showing acoustical directional finding equipment. Drawing by Marek Rys.

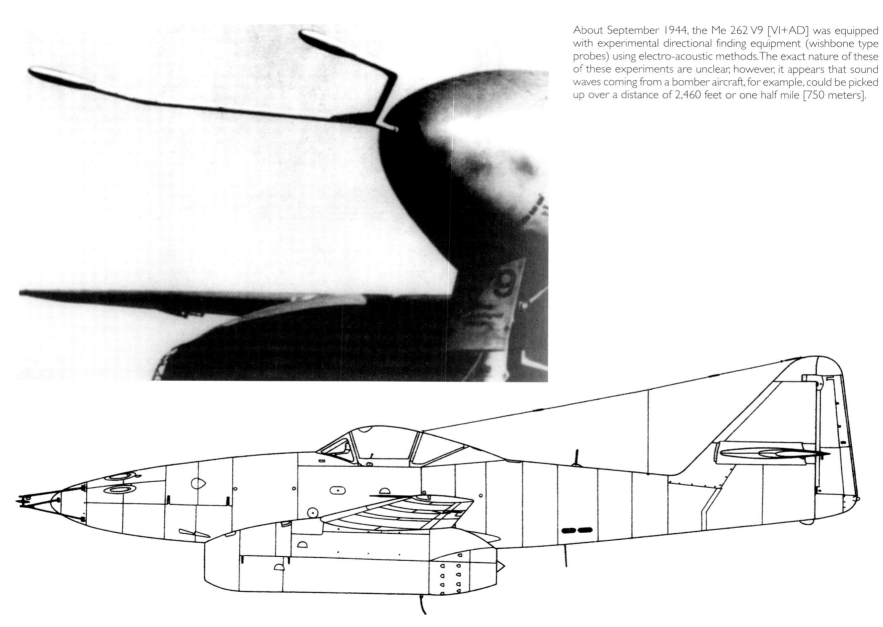

About September 1944, the Me 262 V9 [VI+AD] was equipped with experimental directional finding equipment (wishbone type probes) using electro-acoustic methods. The exact nature of these of these experiments are unclear, however, it appears that sound waves coming from a bomber aircraft, for example, could be picked up over a distance of 2,460 feet or one half mile [750 meters].

In addition to probes installed to pick up sound waves from approaching bombers, the Me 262 V9 [VI+AD] was involved in other similar experiments which are largely unclear. Here the Me 262 V9 has an unusual antenna array attached to its nose cone. Drawing by Marek Rys.

Equipment installed in the nose of the Me 262V9 where the four Rheinmetall-Borsig MK 108 30 mm machine cannon would normally have been. They have been removed for experiments involving the capture of acoustic signals of approaching Allied bombers.

A nose on view of the Me 262V9 with its twin probes installed by Professor Gladenbeck to refine his research into picking up acoustic signals transmitted by Allied bombers. Its cockpit canopy is open to starboard and held open by a single folding metal rod.

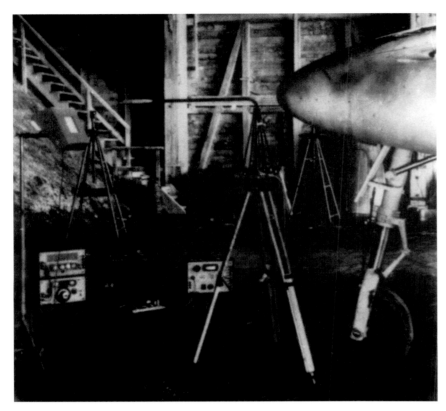

The Me 262 V9 seen here from its port side appears to be involved in electro-acoustic experiments. The nature of these experiments were top secret then and remain unclear today.

The Me 262 V9. Notice the small bushes and tress growing on top of the hangar...an attempt to camouflage.

Looking more like a spy photo, members of Professor Gladenbeck's team mingle around the nose of Me 262 V9...the flying machine used to carry out the Professor's acoustic experiments.

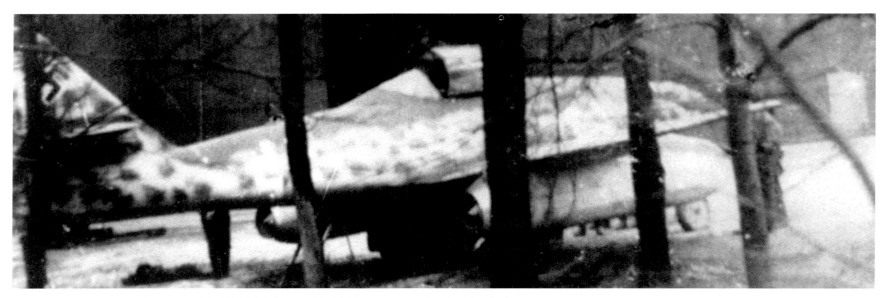

Hidden behind the trunks of pine trees, Me 262 V9 is being used for highly secret acoustic measuring devices.

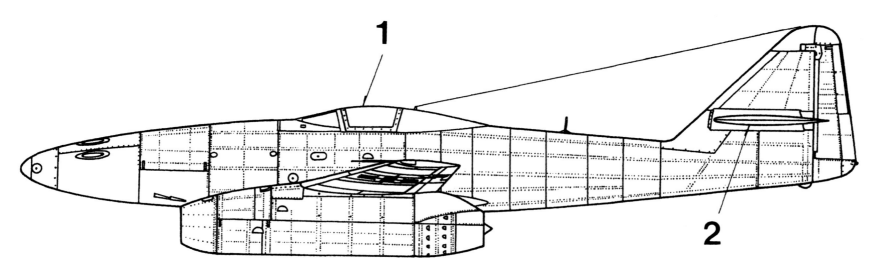

In addition to Me 262 V9's low profile "Rennkabine," or racing cabin (1), Messerschmitt AG designers increased the sweep angle of the horizontal stabilizers (2).

A port side view of Me 262 V9's low profile racing cockpit cabin.

For comparison, this photo shows the Me 262 V9 (VI+AD) seen in flight before its cockpit canopy was modified to the racing cabin. Notice that it still has its production "bubble" canopy.

Me 262 V9 (VI+AD), with its redesigned cockpit canopy, is seen with its nose wheel lifting off the tarmac as seen from its port side.

The Me 262 V9 (VI+AD) featuring its "Rennkabine." The Me 262 V9 was the prototype for the planned high speed Me 262 interceptors.

The Me 262 V9, among others, in the Spring of 1945...photographed in a scrap heap at Lager Lechfeld. The Me 262 V9 is seen on the far left, and its sister flying machine, the Me 262 V10, far right. It is unclear why Me 262 V9's center section fuselage has been painted over in black.

Me 262 V10

- Werknummer 130005;
- Radio call code VI+AE;
- Last prototype;
- Powered by two Junkers Jumo 004B-1 axial flow gas turbine engines;
- First test flight on January 5th 1944;
- Painted full camouflage;
- Equipped with two Rheinmetall-Borsig R-502 solid propellant take off rocket assist units (RATO) to assist the flying machine in the air especially when loaded with an SC 500 kilogram [1, 102 pound] bomb;
- Initially a test bed to investigate stick forces that the flying machine was experiencing during high speed turns. Changes to the ailerons were tested without success. A geared control column which allowed the pilot to apply greater pressure was investigated with success. However, the geared control column was never built into production Me 262 A-1as;
- Damaged during a USAAF bombing raid on Leipheim airfield on February 25th 1944. Repaired to flight status;
- Later a test bed for increased armor protection for the pilot, bomb release mechanisms for its Wikingerschiff bomb racks, RATO packs allowing it to takeoff within 1,968 feet[600 meters], SC 1,000 kilogram [2,204 pound] bombs;
- On May 27th 1944, test pilot Gerd Lindner made the first fight of an Me 262 A-1a with a SC 250 bomb attached to a ETC 503. This was the first Me 262 A-1a on which external bombs were carried! Severe vibrations were experienced. The bomb rack was changed to the Wikingerschiff attached to the fuselage immediately aft the nosewheel well offset to starboard. Assisted by 2xRheinmetall-Borsig R-502 RATO pack the Me 262 V1 was able to take off within 2,000 feet [609.6 meters] and reached a forward speed of 460 miles/hour [74 kilometers/hour]. The test with the Wikingerschiff was successful. The Wikingerschiff was designed by Messerschmitt AG engineers while the ETC 503 was designed by Focke-Wulf Flugzeugbau engineers originally for their Fw 190. The Focke-Wulf designed ETC 503 was modified by Messerschmitt AG engineers to fit on the Me 262 A-1a;
- Bomb-carrying tests showed that the Me 262 V10's superior rate of speed was severely reduced. These losses included: 1. One SC 250 bomb 25 miles/hour [40 kilometers/hour]; 2. Two SC 250 bombs 47 miles/hour [75 kilometers/hour]; 3. One SC 500 bomb 34 miles/hour [55 kilometers/hour];
- Normally the Me 262 A-la's speed was between 515 and 521 miles/hour [830 and 840 kilometers/hour] but with one SC 250 bomb the flying machine's speed was lowered by 62 miles/hour [100 kilometers/hour] reducing the Me 262 A-1a's speed to 460 miles/hour [740 kilometers/hour] thus eliminating its superiority if used as a bomber;
- An effort to increase the range of the Me 262 A-1 a fighter was the so-called "pole-tow" or "Deichselschepp"concept sponsored by the RLM in the fall of 1944, involving one aircraft pulling or towing a winged fuel tank or bomb. It was an effort to increase the bomb load of the Me 262;
- The wing for the towed "flying bomb" came from a Fieseler Fi 103 "buzz bomb;"
- The towing bar was 19 feet [6 meters] in length and a swivel unit was fitted to the tail of the Me 262 V10 which allowed both horizontal and vertical movement of the trailer in flight. Electrically fired explosive bolts were fitted to enable the pilot to jettison the dolly undercarriage after take-off, to release the bomb over the target, and then jettison the trailer, too;
- During tests with a 2,200 pound SC 1000 kilogram bomb considerable "porpoising" of the trailer was experienced, and this movement was transferred via the pole back to the Me 262 V10. It was reported that on one occasion, test pilot Gerd Lindner lost control of his Me 262 and had to bale out;
- During another flight test, a turn by the pilot in the Me 262 created excessive loads on the towing swivel, which tore away from the rear fuselage mount;
- During another flight test the explosive bolts failed to function, but Gerd Lindner skillfully landed the Me 262 with the trailer and bomb still attached;

- With these experiences, the "Deichselschlepp" concept was abandoned, the bomb towing concept being described as "hazardous and unsatisfactory;"
- Flight tests were also tried towing a fuel tank that, when empty, would be uncoupled and landed with the aid of a parachute;
- Tests involved a normal tear drop-shaped drop tank fitted with short rectangular wings and a pair of fixed wheels;
- Another test involving a normal drop tank involved a long wing span of 20.3 feet [6.20 meters] and fitted with a two wheeled jettisonable takeoff trolley/dolly;
- Gerd Lindner found by piloting the "pull-tow" equipped Me 262 V10, October 1944, that the external fuel tank gyrated wildly due to strong buffeting, placing the Me 262 V10 in serious risk;
- The "pull-tow" was found to be too dangerous, although interest in this concept continued right up to war's end;
- Wing damaged on June 8th 1944 by blown tire;
- Believed to have been damaged in February 1945 and never repaired.
- Found post war at Lager Lechfeld;

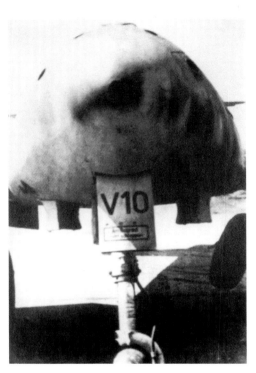

The Me 262 V10, the first Me 262 used in external carrying bomb trials, is seen here in May 1944, with two Messerschmitt AG designed "Wikingerschiff" bomb racks. The "Wikingerschiff" racks created severe air frame vibration due to air turbulence.

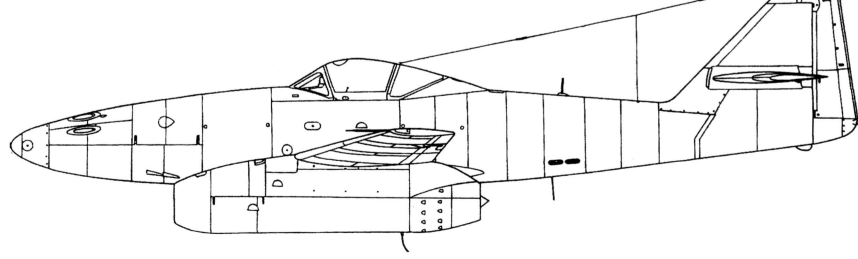

The Me 262 V10 (VI+AE). Illustration by Marek Rys.

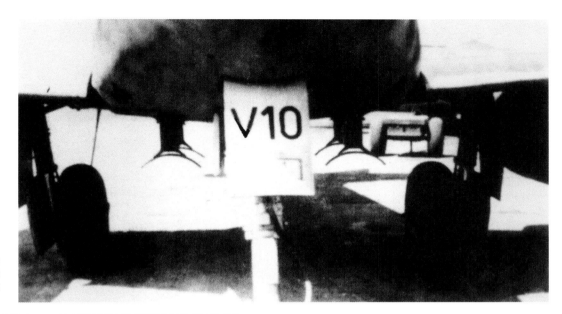

The Me 262 V10 seen in this photograph has been fitted with two Focke-Wulf designed ETC 503 external bomb racks. They created less air turbulence than the Messerschmitt AG "Wikingerschiff" racks so they became the bomb rack of choice.

The Me 262 V10, with a "Wikingerschiff" bomb rack, is being loaded with a SC 500 bomb.

The Me 262 V10 with its SC 500 bomb secured to its "Wikingerschiff" bomb rack. Notice, too, to the far right of the photo is a Rheinmetall-Borsig R-502 solid fuel take off assist rocket.

A nose on view of the Me 262 V10 (Werknummer 130005) seen with two experimental wooden fuel tanks attached to "Wikingerschiff bomb racks. These experiments were disappointing due to the severe air turbulence they created.

A starboard side view of the Me 262 V10 showing it being tested with two experimental wooden external fuel tanks attached to "Wikingerschiff" bomb racks. Notice that the air intakes of the Junkers Jumo 004 gas turbines have wire screens attached.

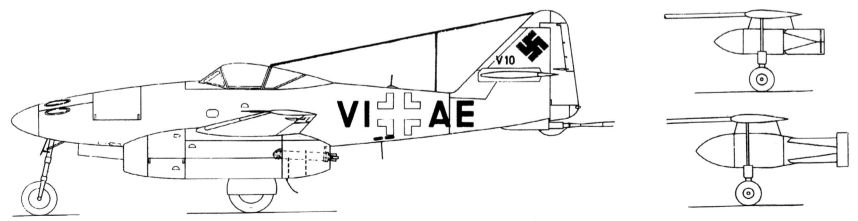

A port side illustration of the Me 262 V10 (VI+AE). On the right top is the towed SC 500 kilogram bomb and right bottom is the towed SC 1000 kilogram [2,204 pound] bomb.

The Me 262 V10 (VI+AE) "Deichselschlepp" with its towed SC 500 kilogram bomb seen moments after lift off.

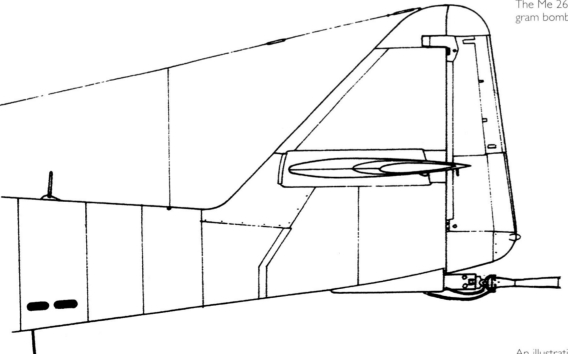

An illustration of the Me 262 V10's tow coupling mechanism beneath the vertical stabilizer and rudder.

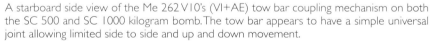

A starboard side view of the Me 262 V10's (VI+AE) tow bar coupling mechanism on both the SC 500 and SC 1000 kilogram bomb. The tow bar appears to have a simple universal joint allowing limited side to side and up and down movement.

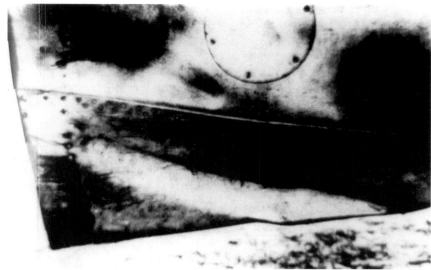

During towed bomb flight trials, the Me 262 V10's coupling mechanism, located below the rudder, was enclosed in a fairing, however, the square black box extended out beyond the fairing.

A close up of the towed SC 1000 kilogram bomb with its tow bar attached to the Me 262 V10. This arrangement is seen from its starboard side. The idea of a pole-towed bomb was quickly discovered to be a dangerous concept for the Me 262 and the idea abandoned.

Testing the towed ordnance called "Deichselschleppverfahren." The Me 262 V10 (VI+AE) is seen shortly after lift off at Lager Lechfeld on October 22nd 1944 towing a SC 1000 kilogram bomb.

The Me 262 V10 with a pole-towed SC 1000 kilogram bomb.

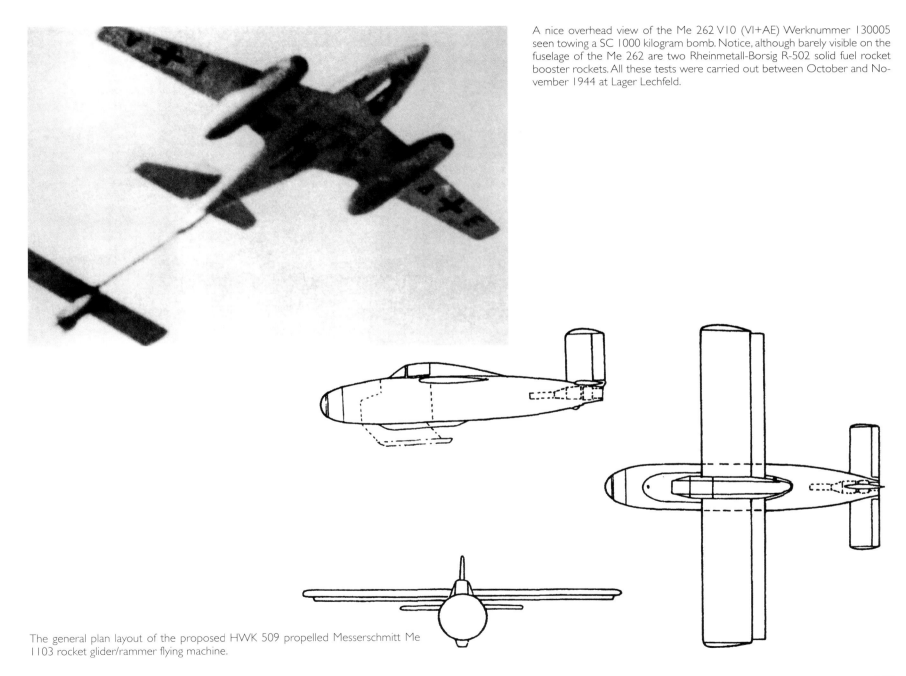

A nice overhead view of the Me 262 V10 (VI+AE) Werknummer 130005 seen towing a SC 1000 kilogram bomb. Notice, although barely visible on the fuselage of the Me 262 are two Rheinmetall-Borsig R-502 solid fuel rocket booster rockets. All these tests were carried out between October and November 1944 at Lager Lechfeld.

The general plan layout of the proposed HWK 509 propelled Messerschmitt Me 1103 rocket glider/rammer flying machine.

The final disposition of Me 262 V10 (VI+AE) at war's end. It sits atop its sister Versuchs machine, the Me 262 V12 (V074) at Lager Lechfeld in the Summer of 1945.

Me 262 V11 [V555]

- Werknummer 110555;
- Known also as White V555 and Me 262 V11;
- Radio call code...unknown;
- Purpose: second prototype designed to perfect a prone bomb aimer laying prone in the flying machine's nose aiming its two bombs via a Lofte 7H gyro-stabilizing bomb sight. The first prototype was Me 262 V8-2 (V484);
- The second prototype V555 was unarmed...with plans to mount two Rhein-metall-Borsig MK 108 30mm machine cannon...their location unclear;
- V555 was fitted with two ETC 504 bomb racks capable of carrying two SC 500 kilogram bombs or two SC 1000 kilogram bombs;
- Due to the increase weight of the bombardier, V555's fuel capacity was reduced to 520 gallons [1,970 liters];
- In January 1945, V555 was fitted with the wooden Lofte Kanzel 11. The nose accommodated a bomb aimer who lay prone using the Lofte 7H gyro-stabilizing bomb sight;

- Flight testing of V555 was performed at Rechlin;
- The main difference between Me 262 V11 (V555) and Me 262 V8-2 (V484) was the installation of long probes fitted each side of the bombardiers cockpit. The purpose of these probes on (V555) is unclear...perhaps for bomb aiming and/or range finding;
- First flight of (V555) was in February 1945, reportedly making 16 test flights between February and March 1945, for a total of flight time of 5 hours 5 minutes;
- On March 30th 1945, V555 was flown by a defecting Luftwaffe pilot to the Allies at Schröck near Marburg. Upon landing, the V555's gear failed to extend and belly landed on its twin Junkers Jumo gas turbines;
- U.S. Army immediately removed both Junkers Jumo 004B-1 gas turbines from V555;
- Final disposition of V555's engineless fuselage is unclear.

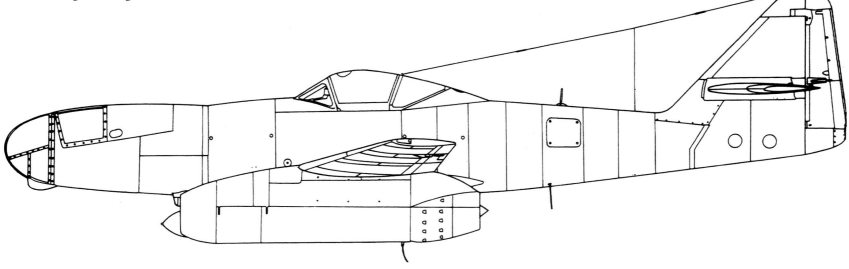

Port side illustration of the Me 262 V11 [V555] by Marek Rys.

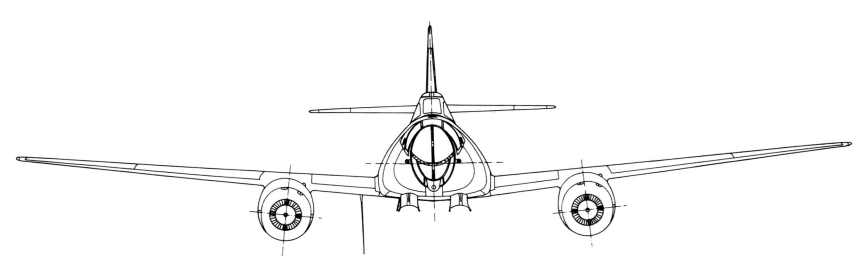

Front-on illustration of the Me 262 V11 [V555] by Marek Rys.

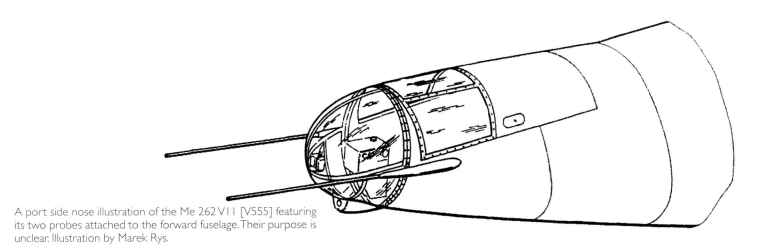

A port side nose illustration of the Me 262 V11 [V555] featuring its two probes attached to the forward fuselage. Their purpose is unclear. Illustration by Marek Rys.

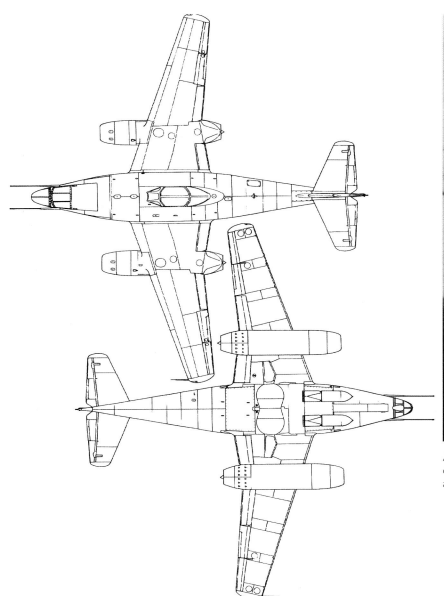

A general upper and lower planform illustration of the Me 262 V11 (V555) by Marek Rys.

An individual opens the bombardier's hatch on the Me 262 V11 (V555). Inside the bombardier was in a prone position looking out the glazed nose. The purpose of the slender probes attached to the fuselage sides just aft the glazed nose is unclear.

The Me 262 VII (V555) seen with full flaps down as it approaches its landing strip after a test flight.

A nice port side view of the Me 262 VII (V555) with its bombardier nose seen resting in the grass on its two Junkers Jumo 004B gas turbines after a defecting Luftwaffe pilot landed it in American occupied Germany. The landing gear failed to operate so the pilot brought it down quite well in the grass with minimal damge.

A starboard nose view of the Me 262 V11 (V555) after a defecting Luftwaffe pilot brought it to American troop occupied Nazi Germany. American soldiers are seen surrounding the prized flying machine. Notice the full glazed nose as well as the upper hatch cover.

Me 262 V12

- Known as the Me 262 V12, or by the last three digits of its Werknummer - (V074);
- Werknummer 170074;
- Known, too, as the Me 262 C-lb "Heimatschützer II;"
- Known, too, as the "Interzeptor ll;"
- Developmentwork started in early 1944, underthe designation Me 262D-1. A single prototype was constructed and the project was eventually redesignated Me 262 C-2b;
- Equipped with two BMW 003R combined gas turbine and bi-fuel liquid rocket engines. The gas turbine was the standard BMW 003A-1 and the bi-fuel liquid rocket engine attached to the 003A-1 was the BMW 109-718 (BMW P-3395);

- The BMW 718 burned a mixture of SV-Stoff (concentrated nitric acid) and R-Stoff (Tonka self-igniting liquid) which provide a thrust of 2,204.6 pounds [1,000 kilograms] for three minutes;
- Purpose of the Me 262 V12 project was to provide a high rate of take off power so that the fighter could reach the Allied bomber formations quickly;
- It was estimated that with the BMW 003R, the Me 262 V12 would be able to reach 39,400 feet [12 kilometers] in 3.9 minutes. This included its take off run;
- Equipped with four Rheinmetall-Borsig MK 108 30mm machine cannon;
- Project Werknummer 170074 did not achieve a single flight before war's end due to an accidental rocket engine explosion during a ground test in January 1945;
- It is unclear if Me 262 V12 was repaired after its fire in January 1945;
- Final disposition at war's end was a scrap heap at Lager Lechfeld;

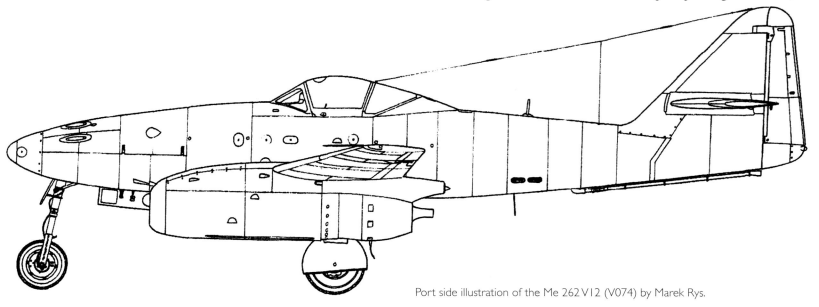

Port side illustration of the Me 262 V12 (V074) by Marek Rys.

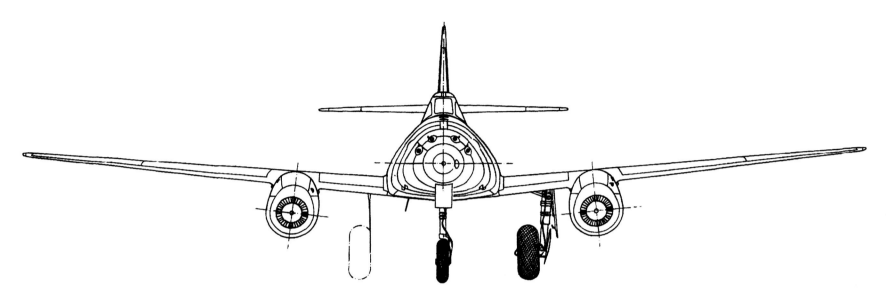

A nose on illustration of the Me 262 V12 (V074) by Marek Rys.

A starboard side view of the Me 262 V12 (V074) by Jamie Davies.

The BMW 003R combined gas turbine and bi-fuel liquid rocket engine. This power plant was developed by Count von Helmuth Zboroski.

A war-time illustration from either BMW or Messerschmitt AG featuring the Me 262 V12 powered by the combined BMW 003R power unit. Notice that the BMW 718 bi-fuel rocket unit is highlighted showing its position beneath the wing's trailing edge

A close up view of the Me 262 V12 (V074) as seen from its starboard side. At least four individuals are attending to the cowlingless starboard BMW 003R power unit.

A static test of the Me 262 V12 (V074)'s port side BMW 718 fitted to the BMW 003R power unit. This successful test occurred on March 23rd 1945, at Lager Lechfeld.

The Me 262 V12 (V074) with its BMW 003R combined gas turbine/bi-fuel liquid rocket engine on fire at Lager Lechfeld Flugplatz in January 1945, after an explosion of the BMW 718 bi-fuel unit.

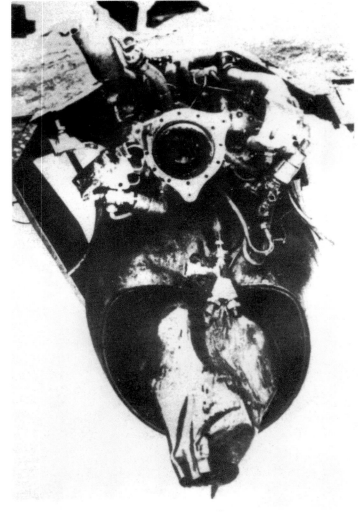

A photo of Me 262 V12 (V074)'s starboard BMW 003R power unit after the explosion and fire of the BMW 718 bi-fuel unit. The BMW 003R is pretty much destroyed, as well as the wing's trailing edge.

The engineless Me 262 V12 (V074) at war's end at Lager Lechfeld Flugplatz aircraft dump. Behind Me 262 V12 is its sister Versuchs prototype the Me 262 V10 VI+AE.

Me 262 Serien "S" Series Prototypes

Due to the rapid destruction of the first batch of twelve Me 262 Versuchs (research) V1 through V12 prototypes from crash landings, severe use, and Allied bombing, quicker than anyone expected, the RLM ordered more prototypes. Up to twenty two of the first production Me 262s were taken off the Messerschmitt AG assembly line at Augsburg and designated "S" or serien meaning series. These twenty two "S" types would be assigned research and development duties. None were given "Versuch" numbers but were identified by their camouflage ... most RLM 76 weissblau or pale blue and designated by a big red numeral on their forward fuselage. Their radio call code appeared in black on the side of the fuselage aft the cockpit cabin. Several carried a small Versuchs number at the base of their vertical stabilizer painted in white. The same for the aircraft's Werknummer. Only the S-1 had a Werknummer and it, too, was painted white. The Balkenhreuz of the B2 and B3 type was painted on both fuselage sides as well as to the underside of the wings. A swastika painted black of the 1-12a type was applied to both sides of the vertical stabilizer. Otherwise there wasn't much else in terms of colors, decorations, or markings on these flying machines.

Power for all the "S" types was provided by Junkers Jumo 004B-1 axial flow gas turbines. Each generally had four Rheinmetall-Borsig MK 108 30mm machine cannon mounted in the nose. Usually they carried a BSK 16 gun camera in the nose cone. A great deal of testing was done to perfect these weapons. Because of centrifugal forces, especially when banking, caused the machine cannon's belt feed mechanism to tear and break.

The tasks asked of the "S" types varied considerably from flying machine to flying machine. Many of these "S" types were handed over to the Me 262 test kommando known as Erprobungskommando 262 and abbreviated EKdo 262 meaning "proving detachment." Other "S" types were flown by Messerschmitt AG test pilots out of Lager Lechfeld.

Just as the initial twelve Me 262 Versuchs prototype were used up, the next batch of twenty two were also consumed in the course of their flight testing and destruction by Allied bombing. Nevertheless, several "S" types achieved outstanding achievements for example Me 262 S-2 when it reached 624 miles/hour [1,004 kilometers/hour]. The twenty S types and their radio call codes, when available, begin with "VI+AF" for Versuchs. The Me 262 S types include the following:

- Me 262 S-1 [VI+AF] Flying tests concentrated on the effectiveness and reliability of its four Rheinmetall-Borsig MK 108 30mm machine cannon.
- Me 262 S-2 [VI+AG] This "S" is one of the most famous of all Versuchs or "S" types. S-2 was the first flying machine in the world to exceed the speed of sound ... 621 miles/hour [1,000 kilometers/hour] piloted by Herlitizius. It reached 624 miles/hour [1,004 kilometers/hours] in a power dive from 23,000 feet altitude. This was its job ... high speed testing/development and the operation of its four Rheinmetall-Borsig MK 108 30mm machine cannon.
- Me 262 S-3 [VI+AH] Used by the Luftwaffe group Erprobungskommando Thierfelder to test mechanical improvements made in the operation of the four Rheinmetall-Borsig MK 108 30mm machine cannon. It was also used to evaluate improvements made to its landing gear especially nosewheel development and testing.
- Me 262 S-4 [VI+AI] Used by the Luftwaffe group Erprobungskommando Thierfelder to test modifications made in its flight controls such as the installation of a Flettner counterbalance on its horizontal stabilizer.
- Me 262 S-5 [VI+AJ] This is another significant "S" type in that it would become the prototype for the tandem two seat pilot trainer designated Me 262 B-1. It was modified by Blohm & Voss-Wenzendorf and delivered to the Luftwaffe Test Center Rechlin/Lars in July 1944.
- Me 262 S-6 [VI+AK] No specific flight test duties, however, it appears that it was flown by members of Erprobungskommando 262 to gain flying proficiency in the gas turbine powered Me 262.
- Me 262 S-7 [VI+AL] Specific flight duties unclear, however, it appears that it was, too, used by pilots of Erprobungskommando 262, under the command of Hauptmann Werner Thierfelder, to gain flying proficiency in the gas turbine powered Me 262. It crashed due to a gas turbine fire on June 1 st 1944, never flew again, and was used for landing gear tests.

- Me 262 S-8 [VI+AM] Specific flight testing duties unclear.
- Me 262 S-9 [VI+AN] Specific flight testing duties unclear.
- Me 262 S-10 [V11+AO] Flight duties included the testing of wheel brake modifications, wooden tail assembly, wooden control surfaces, aileron modifications, leading edge wing slots, and a adjustable control column.
- Me 262 S-11 [V11+AP] Specific flight testing duties unclear.
- Me 262 S-12 [VI+AQ] Believed to be the first Me 262 to shoot down an Allied flying machine (an RAF Mosquito) by Hauptimann Werner Thierfelder of Erprobungskommando 262 on July 26th 1944.
- Me 262 S-13 [VI+AR] Specific flight testing duties unclear, however, it was assigned to Erprobungsstelle Rechlin Test Center.
- Me 262 S-14 [VI+AS] Believed to have been flown by pilots of Kampfgeschwader 51 (I/KG 51) or fighter/bombers, and if so, specific flight duties unclear. It would be I/KG 51 to suffer the first loss of an Me 262 A-1a on August 8th 1944, by two Republic P-47s from the USAAF 8th Fighter Squadron.
- Me 262 S-15 [VI+AT] Specific flight testing duties unclear, however, assigned to Erprobungskommando 262.

Serien preproduction types Me 262 S-16 through Me 262 S-22 and their flight testing duties are not clear. In fact, no photographic evidence is known to this author regarding the specific flight testing duties of S-16 through S-22.

Me 262 S-1

- The Me 262 S-1 is the first of seven Me 262 aircraft produced when series production of the Me 262 started in. It is reported that as many as twenty-two of the first Me 262s manufactured in series production may have been taken for use as designated flight test machines. This was required because so many of the designated "V pre-production test machines had been destroyed during flighttesting. All "S" machines were powered by the typical Junkers Jumo 004B-1 axial flow gas turbine. Like the pre-production "V machines before them, most of these "S" flying machines were usually consumed, too, as a result of their severe flight testing duties ... many of which were later written off as suffering damage beyond repair. The first of these Serien or series mules is "S-1." Its Werknummer was 130006, and it carried radio call code "VI+AF" on its fuselage and wing undersurface in large black letters. A large numeral " 1 " was painted on its forward fuselage side in red. Overall, S-1 was painted in RLM 76 "weissblau," or pale blue gray;
- S-1 stands out from its sister "S-machines" because it had a full functioning tail light built into the tail section beneath its rudder;
- First flight of the S-1 occurred on April 19" 1944. Its pilot was Oberleutant Ernst Tesch of the Luftwaffe's designated Me 262 test kommando known as Erprobungskommando 262 or "EKdo 262." S-1 was officially signed over to EKdo 262 on April 25" 1944. Subsequent testing concentrated on the effectiveness and reliability of the four Rheinmetall-Borsig MK 108 30mm machine cannon installed in its fuselage nose. The "S-1" also carried a fuselage nose-mounted BSK 16 gun camera;
- April 27th 1944. Suffered wing and fuselage damage due to a main wheel tire blow out during a take run. Repaired to flight status;
- June 11th 1944. Starboard wing severely damaged due to hard landing. Repaired to flight status;
- June 16th 1944. Written off when it was severely damaged in a crash landing, approximately two months after its first test flight. Its crash, due to pilot error by Oberfeldwebel Becker, came when he failed to have the landing fully down/extended priorto touch down on grass. The S-1's main wheels and gas turbines were torn from their mountings. Written off as unrepairable;

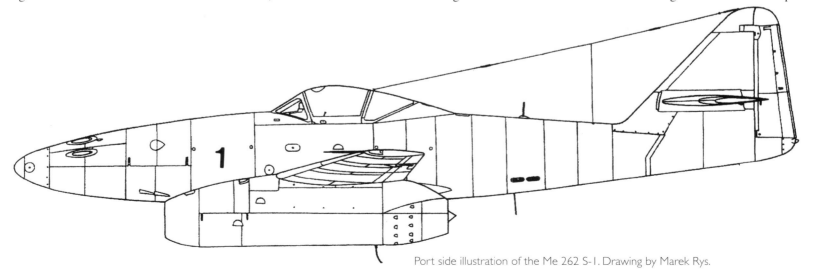

Port side illustration of the Me 262 S-1. Drawing by Marek Rys.

The Me 262 S-1. Courtesy: Me 262 Volume One, J. RIchard Smith and Eddie J. Creek, Classic Publications, Crowborough, England, 1997.

The first pre-production Me 262 "S" flying machine...the "S-1." It was painted in a pale blue camouflage known as RLM 76 "weissblau."

Me 262 S-1 Werknummer 130006 was assembled at Messerschmitt AG-Augsburg. This photo shows the machine listing to starboard due to tire blowing out on its take off run April 27th 1944.

The Me 262 S-1's aft cockpit as seen from its port side. There does not appear to be a pilot's head cushion in this early "S" type.

Me 262 S-1's seen undoing major maintenance and/or modifications. Notice that its nose cone has been removed as well as the fuselage hatch cover over its four Rheinmetall-Borsig MK 108 30mm machine cannon. The Me 262 S-1's first flight was on April 19th 1944.

The Me 262 S-1. A mechanic is pointing to a strengthening plate attached to the aft fuselage. The need for this modification is unclear. S-1's radio call code(VI+AF) was painted in black. Notice that its Werknummer (130006) is painted white and located directly beneath the horizontal stabilizer.

Me 262 S-2

- This Me 262 version is one of the most famous of all...perhaps the most significant flying machines of World War Two. S-2 reportedly was the first flying machine in the world to exceed the speed of sound ... 621 miles/hour [1,000 kilometers/hour]. This achievement occurred on June 28th 1944 when pilot Feldwebel Heinz Herlitizius of EKdo 262 put the S-2 in a 35 degree power dive from a height of 22,966 feet [7,000 meters] and went on to reach 624 miles/hour [1,004 kilometers/hour]. Heinz Herlitizius survived the war only later to lose his life post war in an automobile accident;
- Werknummer 130007, however, this number does not appear on the aircraft;

- Painted in RLM 76 "weissblau "pale blue gray with its series number "2" appearing on the forward fuselage side. It is painted red;
- Radio call code VI+AG painted black;
- First flight occurred on March 28th 1944;
- Flight testing duties involved high speed tests and the operation of its four Rheinmetall-Borsig MK 108 30mm machine cannon and its nosemounted BSK 16 gun camera. S-2 was later converted to carry out photo aerial reconnaissance testing;
- S-2 was destroyed on the ground on July 1 9th 1944 from an Allied bombing raid. At the time of its destruction it had completed 47 test flights and logged over twenty hours of flight time;

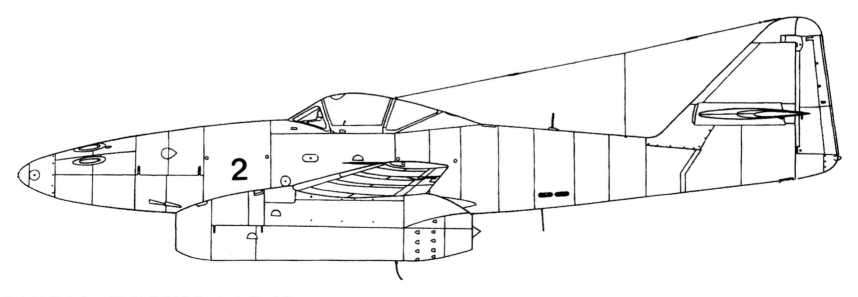

Port side illustration of the Me 262 S-2. Drawing by Marek Rys.

Me 262 S-3

- Werknummer 130008.
- Radio call code VI+AH.
- Powered by twin Junkers Jumo 004B-1 axial flow gas turbine engines.
- Painted overall in RLM 76 "weissblau" or pale gray with its radio call code black and its series number "Y red and appearing on the forward fuselage side.
- Me 262 S-3's first flight occurred on April 5th 1944.

- Me 262 S-3 was used to test modifications made in the four nosemounted Rheinmetall-Borsig MK 108 30mm machine cannon and its BSK 16 nose cone mounted gun camera. This testing was carried out by EKdo 262-Lager Lechfeld Flugplatz.. Messerschmitt AG also used Me 262 S-3 to test mod - ifications/improvements made to the Me 262s tricycle landing gear hydraulics.
- Me 262 S-3 was severely damaged on June 16 th 1944, during a test flight at EKdo 262-Lager Lechfeld due to pilot error. It appears that the pilot was late in extending the landing gear and when he touched down the gear was not fully extended and locked. S-3 was written off as unrepairable.

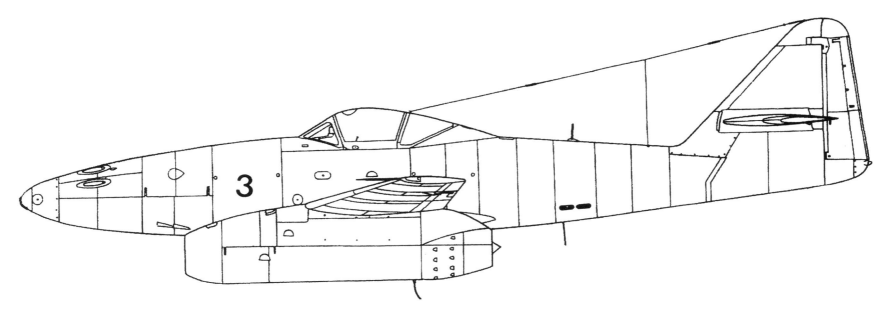

Port side illustration of the Me 262 S-3. Drawing by Marek Rys.

Me 262 S-3 after a very hard landing at Lager Lechfeld Flugplatz on June 16ᵗʰ 1944.

A close up of Me 262 S-3 with its complete nose assembly down in the grass after a failed nose gear brought it down.

Above: Pieces of Me 262 S-3 appear to be lying everywhere...the port side Junkers Jumo several yards back, the starboard gas turbine lays under the fuselage, undercarriage, and cowling. It would have taken a great deal of energy to rip a gas turbine from its mounting as shown in the photo. So it is not clear what happen here.

Left: Me 262 S-3's port side gas turbine after being ripped from its mounting attachments under the wing.

Me 262 S-4

- Werknummer 130009.
- Radio call code VI+Al.
- First flight of Me 262 S-4 [VI+Al] is reported to occurred sometime in early May 1944.

- S-4 was used to flight test, among other duties, modification in its flight controls with the addition of a Flettner counterbalance on its horizontal stabilizer.
- Destroyed on the ground at Lager Lechfeld Flugplatz on July 19th 1944 during a USAAF bombing raid.

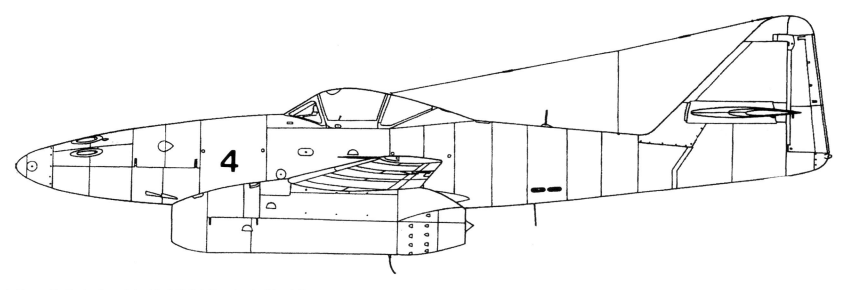

Port side profile illustration of the Me 262 S-4. Drawing by Marek Rys.

Me 262 S-3 Werknummer 130009 had its first flight on May 1st 1944, then this accident a few weeks later which tore away its port side leading edge wing slots. Notice the fitted canvas cover over the port side gas turbine's exhaust.

Me 262 S-5

- Werknummer 130010.
- Radio call code VI+AJ.
- Powered by twin Junkers Jumo 004B-1 axial flow gas turbine engines.
- Painted overall in RLM 76 "weissblau" or pale gray with its radio call code black and its series number "S" in red and appearing on the forward fuselage side.

- S-5 is a significant version because it became the prototype for the tandem two-seat pilot trainer Me 262 B-1. Conversion of the single seat Me 262 S-5 to the prototype Me 262 B-1 was performed by Blohm & Voss Flugzeugbau.
- The Blohm & Voss modified Me 262 S-5 was severely damaged on its 31st flight on October 8th 1944. Based on surviving photographs, S-5 appeared to have suffered a landing gear collapse upon landing and flipping over killing its test pilot Stabsing Heinz Pfister and a student pilot by the name of Rodloff. Me 262 S-5 was written off as unrepairable. It has logged over 8 hours of flight time.

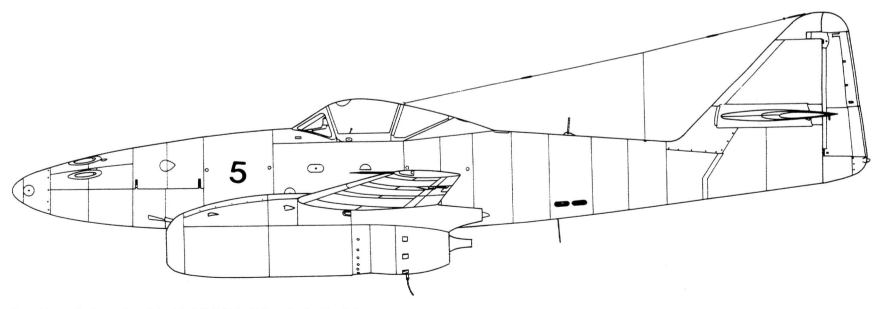

Port side profile illustration of the Me 262 S-5 VI+AJ. Drawing by Marek Rys.

The Blohm & Voss Flugzeugbau two man night fighter prototype, which they modified from the single seat S-5. Notice the white lighting bolt on the forward fuselage.

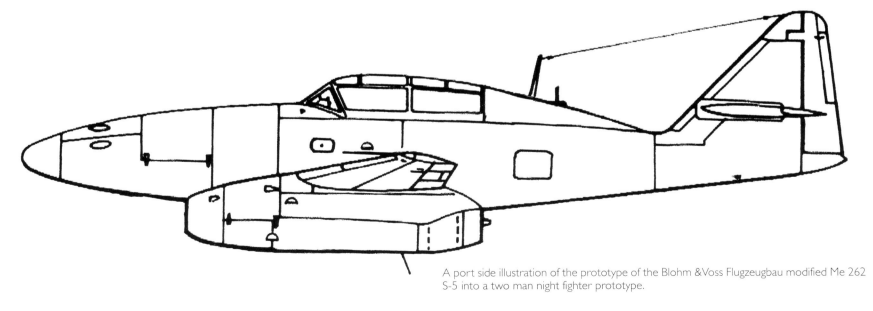

A port side illustration of the prototype of the Blohm & Voss Flugzeugbau modified Me 262 S-5 into a two man night fighter prototype.

Me 262 S-6

Luftwaffe Hauptmann (Captain) Werner Thierfelder (1915-1944).

- Werknummer 130011.
- Radio call code VI+AK.
- Delivered to Erprobungskommando 262 [EKdo 262] immediately after its assembly sometime in May 1944. S-6 does not appear to have been used for any specific flight test duties purposes/duties, however, S-6 was flown extensively by members of EKdo 262 to gain flight proficiency in the Me 262.
- First flight of Me 262 S-6 [VI+AK] is reported to occurred sometime in May 1944.
- S-6 was completely destroyed on July 18th 1944, killing its skilled pilot and EKdo 262's commander Hauptmann (Captain) Werner Thierfelder when the aircraft dove into the ground at high speed. It is reported that Thierfelder was in pursuit of Allied aircraft, or perhaps being pursued himself, and put the S-6 into a high speed dive from which he could not recover.

Port side illustration of the Me 262 S-6. Drawing by Marek Rys.

Me 262 S-7

- Werknummer 130012, however, this number does not appear anywhere on the aircraft's fuselage as it did on several other "S" types.
- Radio call code VI+AL.
- Delivered to Erprobungskommando 262 [EKdo 262].
- Although Me 262 S-7 was the seventh preproduction Me 262 "S" series, it designated S-6 with a large red "6" painted on its forward nose.
- It is not clear, what if any flight testing duties, S-6 participated in while with EKdo 262. This preproduction "S" machine may have been used solely to provide EKdo 262 pilots with the opportunity to gain proficiency in piloting the gas turbine powered Me 262.
- First flight of Me 262 S-7 [VI+AL] is reported to occurred sometime in April 1944.
- S-6 was severely damaged (damaged nose, wings, and both gas turbines engines) on June 1st 1944, while attempting a landing with one gas turbine on fire. Written off as unrepairable, however, the air frame was reportedly used for undercarriage modifications and retraction and extension tests.
- Final disposition at war's end on May 8th 1945, is unclear.

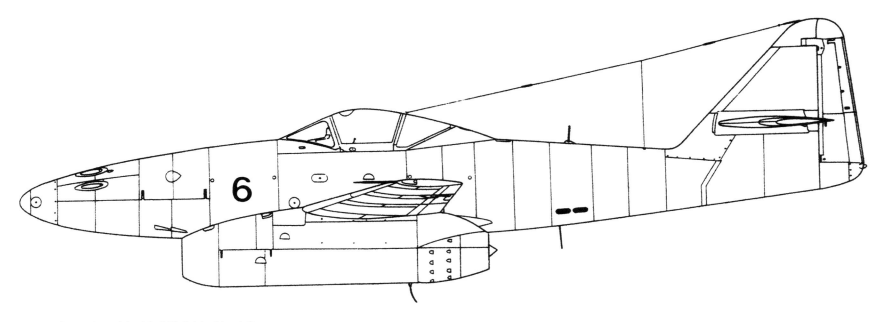

Port side illustration of the Me 262 S-6 by Marek Rys.

The June 1st 1944 hard landing of Me 262 S-7 (VI+AL) Werknummer 130012 due to gas turbine fire. This flying machine belonged to the Erprobungskommando 262 Test Center. Although "6" is painted on the its fuselage nose, this was really the "7th" in the "S" series. Notice how "S-7's" port gas turbine is tucked under the fuselage. In the background to the right of the photo can be seen a lifting crane needed to salvage the downed S-7.

A nose port side view of Erprobungs-kommando 262's downed Me 262 S-7 which had been given the S-series number of six.

Me 262 S-8

- Werknummer 130013.
- Radio call code VI+AM.
- Date manufactured and location ... unclear.
- Test flight unit VI+AM given over/assigned ... unclear.

- First flight sometime in May 1943 ... date unclear.
- Specific air frame modifications and experimental flight test duties unclear.
- Reportedly destroyed on July 19th 1944, due to a USAAF bomber raid location at the time of its destruction is unclear.

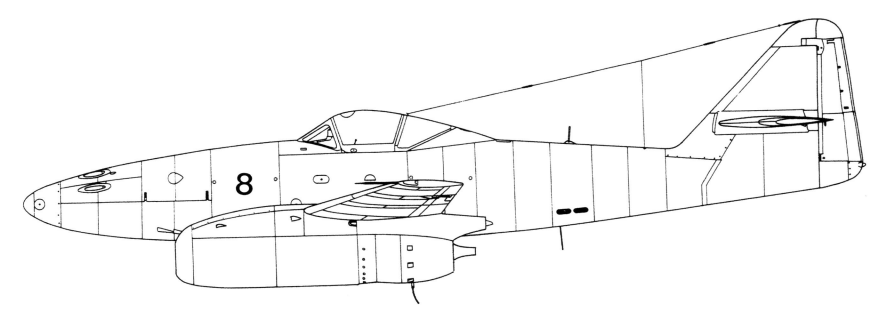

Port side illustration of the Me 262 S-8 by Marek Rys.

Me 262 S-9

- Werknummer 130014.
- Radio call code VI+AN.
- Date manufactured and location ... unclear.

- Test flight unit VI+AN given over/assigned ... unclear.
- First flight sometime in May 1944 ... date unclear.
- Specific air frame modifications and experimental flight test duties unclear.
- Final disposition at war's end in Europe is unclear.

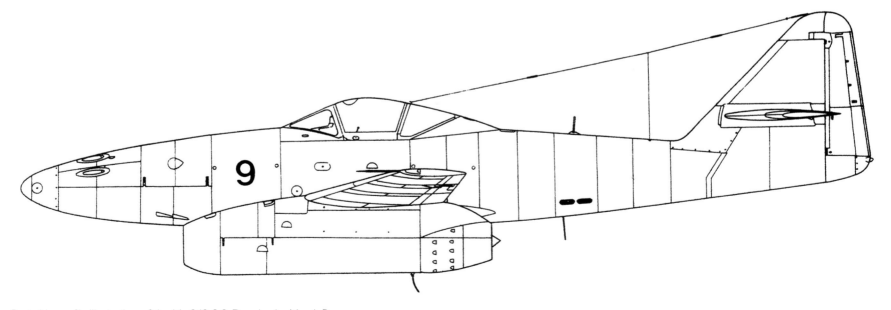

Port side profile illustration of the Me 262 S-9. Drawing by Marek Rys.

Me 262 S-10

- Werknummer 130015.
- Radio call code VI+AO.
- Initially designated Me 262 S-10, however, reportedly known also as Me 262 V1 or Me 262 V1-2.
- Date manufactured and location ... unclear.
- First flight June 30th 1944.
- Reported to have completed 117 test flights totaling over 40 hours of flight time.

- Air frame modified to experiment with leading edge wing slots, main wheel main changes, and an adjustable pilot's control column/stick. Me 262 S-10 is reported to have had an experimental tail assembly with wooden control surfaces.
- After its use as a test flying machine, Me 262 S-10 is believed to have been transferred to Jagdverband JV44, the so-called "Galland Circus." This, too, is unclear.
- Final disposition of Me 262 S-10 [V[+AO] at war's end in Europe is un clear.

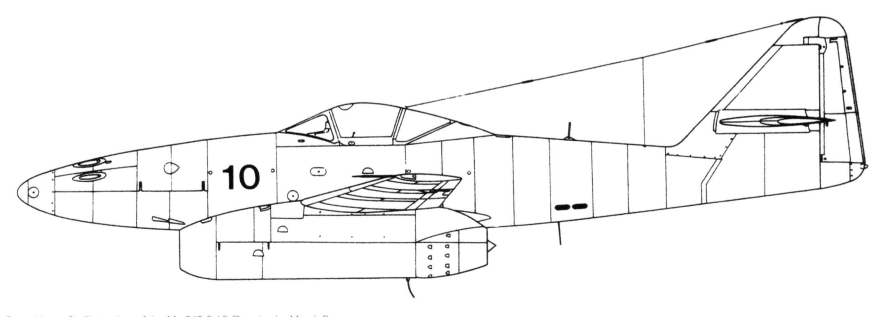

Port side profile illustration of the Me 262 S-10. Drawing by Marek Rys.

Me 262 S-11

- Werknummer 130016.
- Radio call code VI+AP.
- Unclear who flight tested Me 262 S-11 and use/duties.

- First flight of Me 262 S-11 [VI+AP] is reported to occurred sometime in May 1944.
- Final disposition at war's end on May 8th 1945, is unclear.

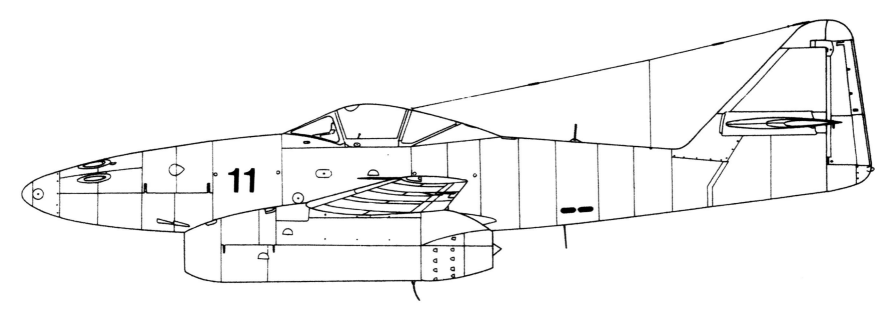

Port side profile illustration of the Me 262 S-11. Drawing by Marek Rys.

Me 262 S-12

- Werknummer 130017.
- Radio call code VI+AQ.
- Delivered new to fighter unit Erprobungskommando 262 (EKdo 262) Kommando Thierfelder.

- On July 26th 1944, Leutnant (Lieutenant) Alfred "Bubi" Schreiber shot down an RAF Mosquito reconnaissance aircraft. It is the first aerial victory by a Me 262 in WWII. Leutnant Schreiber lost his life on November 26th 1944, when his Me 262 crashed near Lager Lechfeld for reasons which are un clear.
- VI+AQ was reportedly written off in October 1944, for reasons which are unclear.

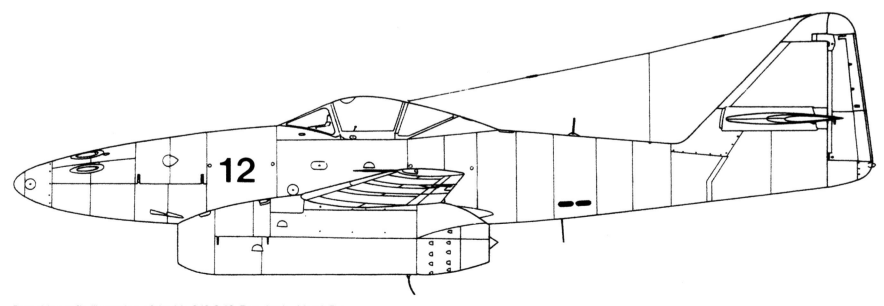

Port side profile illustration of the Me 262 S-12. Drawing by Marek Rys.

Leutenant Alfred "Bubi" Schreiber...the first Luftwaffe pilot to score a "kill" with a Me 262 fighter in WWII.

A de Havilland "Mosquito" of the type shot down by Leutenant Schreiber and making history as the first "kill" of the war for a Me 262 (S-12 VI+AQ).

Me 262 S-12 [Werknummer 130007] flown by Leutenant Schreiber when he shot down a RAF "Mosquito" and scoring the first kill by an Me 262 of an Allied flying machine.

Me 262 S-13

- Werknummer 1300018.
- Radio call code VI+AR.
- Delivered new to Erprobungsstelle Test Center - Rechlin with radio call code VI+AR.
- After arriving at Rechlin, S-1 3 [VI+AR]'s radio call code was changed to E3+01 meaning Department E3 - Engines at Erprobungsstelle Test Center - Rechlin.

- First flight of Me 262 S-13 was sometime in June 1944.
- As E3+01 (S-13), in conjunction with specialists from Junkers Jumo, was used for rapid throttle movement tests at high altitude with the new double gas turbine fuel regulator, engine thrust measurements, engine acceleration behavior and duration tests.
- Written off at Erprobungsstelle Rechlin sometime in October 1944, for reasons which are unclear.

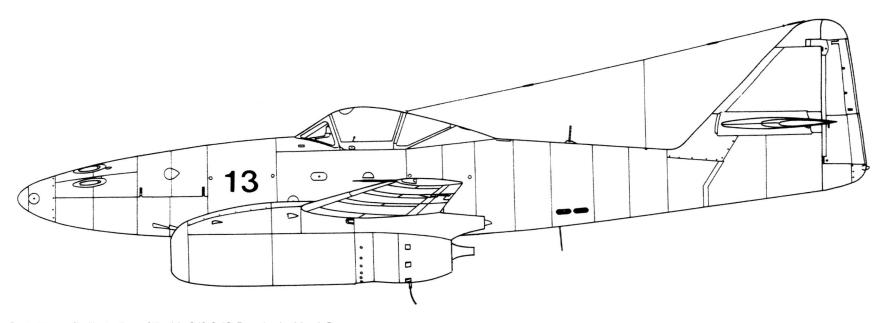

Port side profile illustration of the Me 262 S-13. Drawing by Marek Rys.

Port side view of the Me 262 S-13 VI+AR appearing being ready for a test flight.

Me 262 S-14

- Werknummer 130019.
- Radio call code VI+AS.
- Date assembled and location is unclear.
- First flight sometime in June 1944.

- Reported to have been assigned to Gefechtverband (combat formation) I/ KG 51 and later known as NSGr 20 used in attacking ground Allied targets usually airfields and troop concentrations in late 1944, using armor piercing bombs and anti-personnel (fragmentation) bombs.
- Final disposition of Me 262 S-14 [VI+AS] at war's end is unclear.

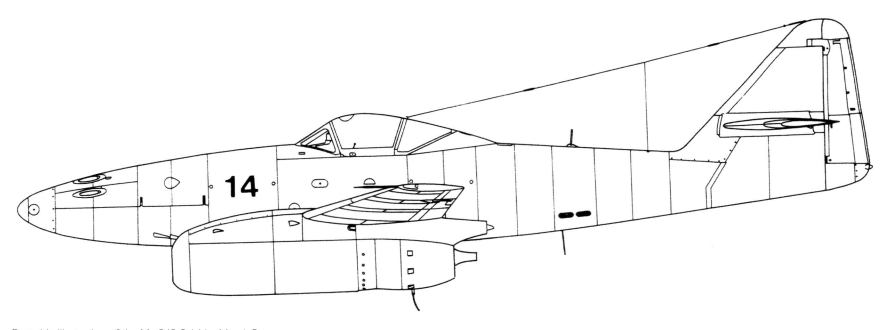

Port side illustration of the Me 262 S-14 by Marek Rys.

Me 262 S-15

- Werknummer 130020.
- Radio call code Vl+AT.
- Date assembled and location is unclear.
- First flight sometime in June 1944.

- Reported to have been assigned to Erprobungskommando 262 (EKdo 262). Specified uses and flight test duties with EKdo 262, if any other than flight proficiency training for its pilots, are unclear.
- Final disposition of Me 262 S-15 [VI+AT] at war's end in Europe is unclear.

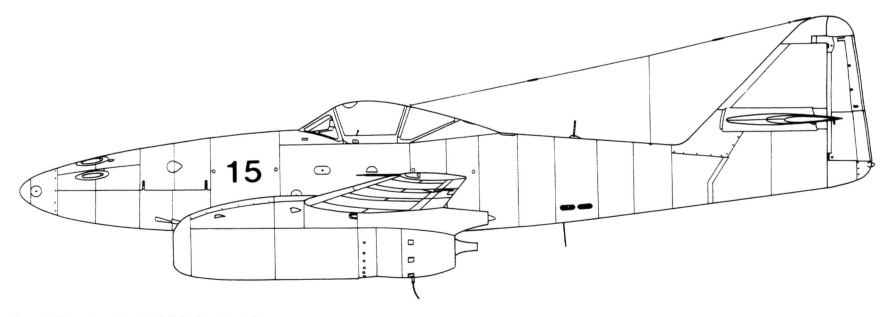

Port side illustration of the Me 262 S-15 by Marek Rys.

Me 262 S-16 to S-22

- At least seven additional "S" type pre-production Me 262s are known to have been assembled and delivered to Erprobungsstelle Test Center Rechlin and Erprobungskommando 262. Some were used for additional flight testing while others were assigned combat duties such as Me 262 S-1 2 and S-14. However, the details of these seven "S" pre-production types are unclear.

One of the last Me 262 "S" pre-production flight test machines. With the end of the "S" types serial production of the Me 262 A-1a began.

Color Profiles
Willy Messerschmitt's Hand-drawn Sketch of October 17ᵗʰ 1939
for a Gas Turbine-powered Pursuit Fighter and a Tricycle Landing Gear

Digital image of Willy Messerschmitt's hand drawn sketch by Jozef Gatial.

Messerschmitt Project Design 65
Straight Low Wing with Oval Fuselage. BMW P-3302 Gas Turbines through the Wing

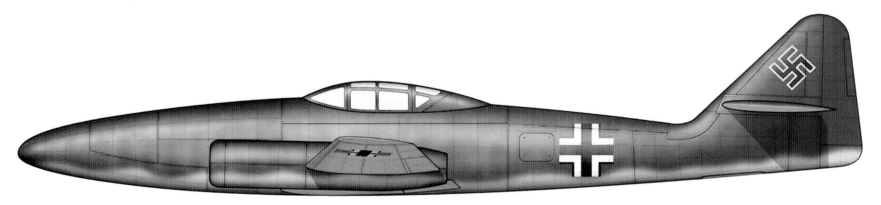

Digital image of Messerschmitt Project Design with a straight low wing, oval fuselage, and BMW P-3302 gas turbines through the wing. by Jozef Gatial.

Messerschmitt Project Design 65
Straight Mid Wing with Oval Fuselage. BMW P-3302 Gas Turbines through the Wing

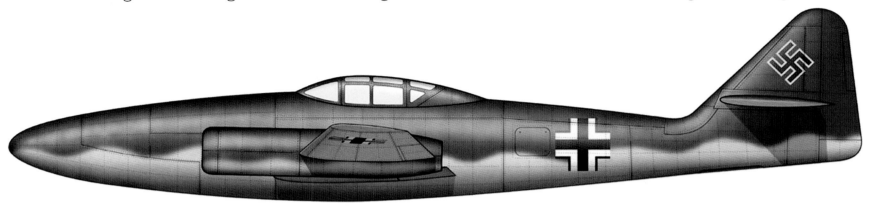

Digital image of the Messerschmitt Project Design 65 with a straight mid wing, oval fuselage, and BMW P-3302 gas turbines through the wing. Jozef Gatial.

Messerschmitt Project Design 65
Wind Tunnel Drawing without Gas Turbines by AVA, July 14th 1940

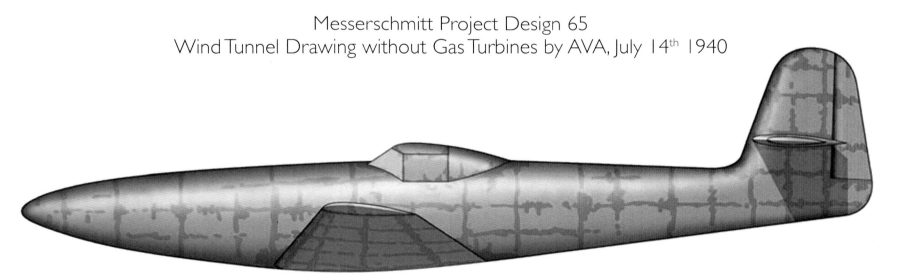

A port side profile of Woldemar Voigt's evolving Me P-65. With this version, from June 7th 1939, Voigt's team was closing in on a working design suited for mass production and what will eventually be known as the Me 262 A-1a. Digital image by Jozef Gatial.

Messerschmitt Project Design 65
AVA Wind Tunnel Model with BMW P-3302 Gas Turbines
Mounted above the Wing's Surface

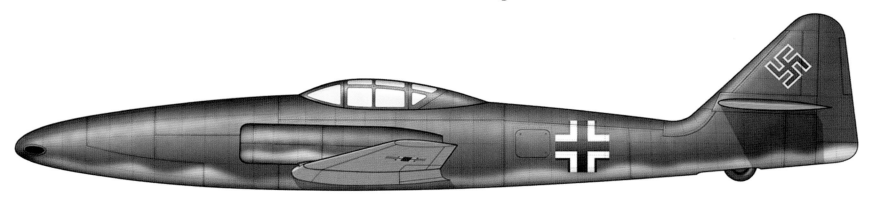

Port side profile of Me P-65 featuring BMW P-3302 gas turbines mounted above the wing. AVA wind tunnel model. Digital image by Jozef Gatial.

Messerschmitt Project Design 65
AVA Wind Tunnel Model with BMW P-3302 Gas Turbines
Mounted high above the Wing's Surface

Port side profile of Me P-65 featuring BMW P-3302 gas turbines mounted high above the wing's surface. AVA wind tunnel model. Digital image by Jozef Gatial.

Messerschmitt Project Design 65
Low Wing Straight Inner Wing with 18° Swept Outer Wing
Triangular Fuselage BMW P-3302 Gas Turbines through the Wing Spar

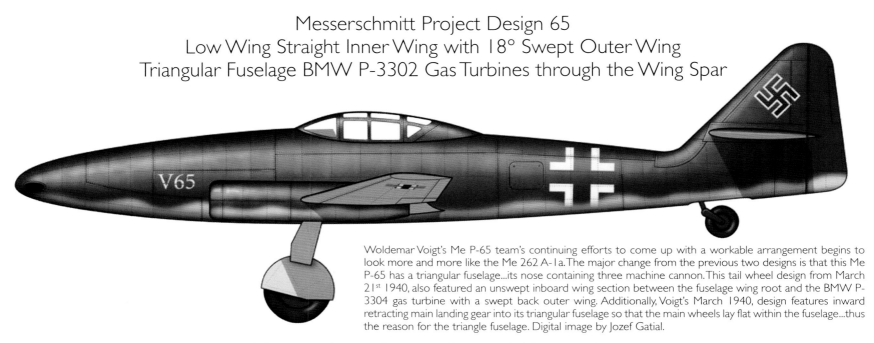

Woldemar Voigt's Me P-65 team's continuing efforts to come up with a workable arrangement begins to look more and more like the Me 262 A-1a. The major change from the previous two designs is that this Me P-65 has a triangular fuselage...its nose containing three machine cannon. This tail wheel design from March 21st 1940, also featured an unswept inboard wing section between the fuselage wing root and the BMW P-3304 gas turbine with a swept back outer wing. Additionally, Voigt's March 1940, design features inward retracting main landing gear into its triangular fuselage so that the main wheels lay flat within the fuselage...thus the reason for the triangle fuselage. Digital image by Jozef Gatial.

Messerschmitt Project Design 262 Modell I
Straight Low Wing Triangular Fuselage BMW P-3302 Gas Turbines below the Wing

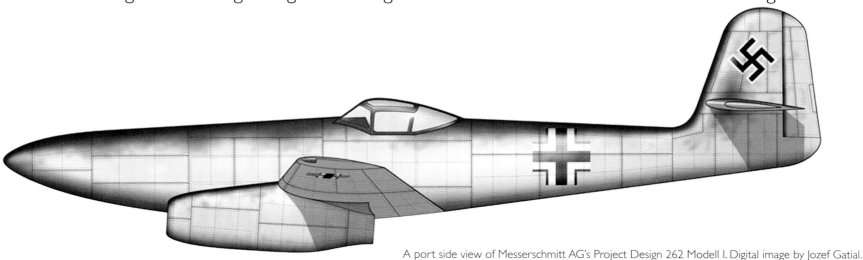

A port side view of Messerschmitt AG's Project Design 262 Modell I. Digital image by Jozef Gatial.

Messerschmitt Project Design 262 Evolving Post Modell I
Full 35° Sweep Low Wing AVA Wind Tunnel Model

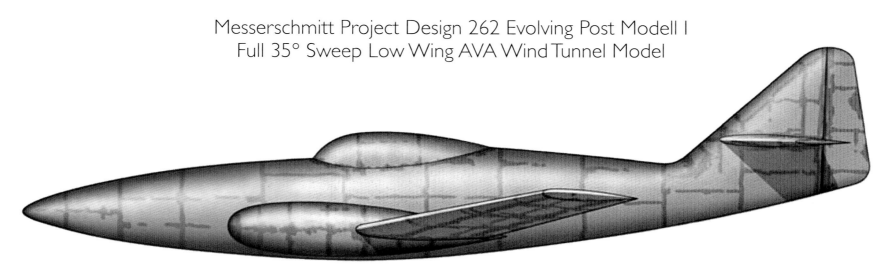

The Me 262 - post Modell I from 1940 by AVA to test the effects of its 35 degree wing sweep. Digital image by Jozef Gatial.

Messerschmitt Project Design P-262
Featuring (for the First time) a Nose Wheel

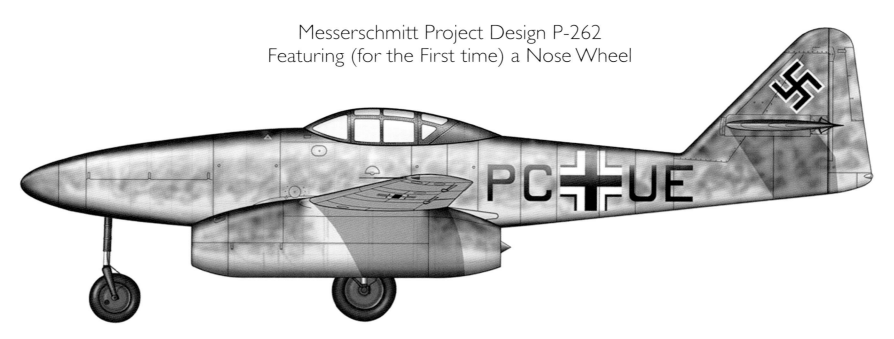

Me 262 project design shown for the first time with a nose wheel. Digital image by Jozef Gatial.

Messerschmitt Project Design 1070
Swept Back Low Wing Oval Fuselage Nose Wheel, BMW P-3302 Gas Turbines through the Wing

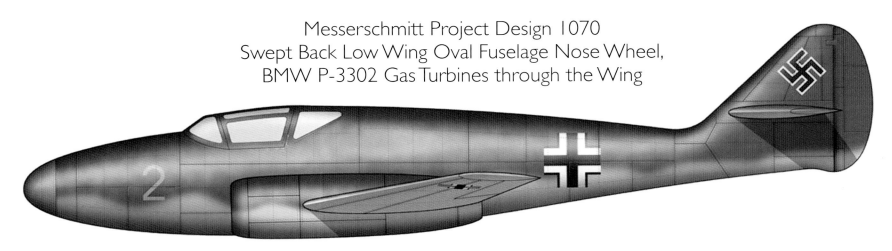

Port side profile of the Me P-1070 "pursuit fighter" featuring its gas turbines mounted through the wing. Digital image by Jozef Gatial.

Me 262 V1 Stage 1

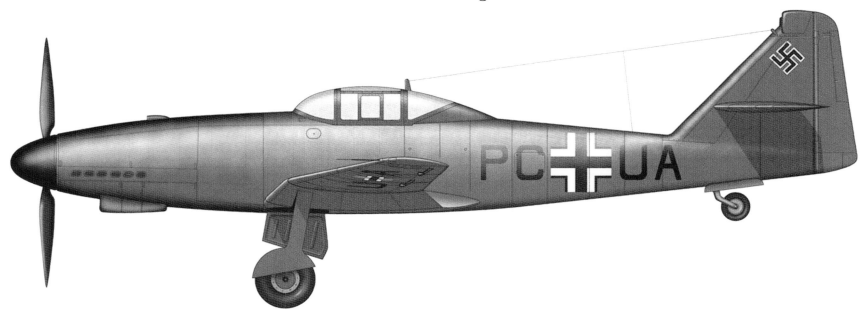

Port side fuselage profile of the propeller-driven Me 262 V1 Stage 1. Digital image by Jozef Gatial.

214

Me 262 V1 Stage 2

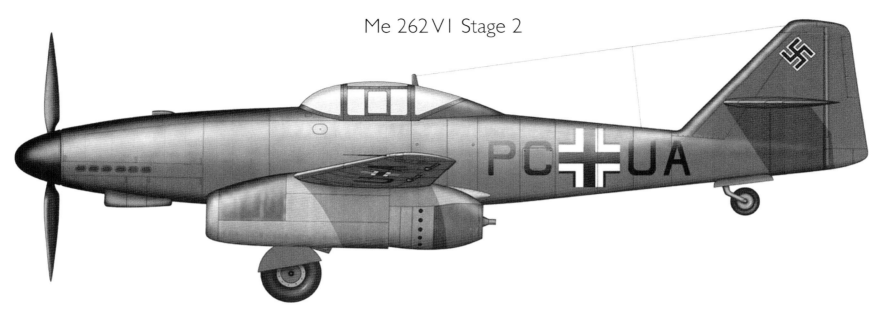

Port side fuselage view of the hybrid (one piston engine and two gas turbines) Me 262 V1 Stage 2. Digital image by Jozef Gatial.

The hybrid motored Me 262 V1 - Stage 2. It is shown immediately after lift off as its pilot Fritz Wendel spoke in the radio..."I've got one turning and two burning." Digital image by Mario Merino.

The hybrid Me 262 V1 - Stage 2 seen leaving the runway behind. Digital image by Mario Merino.

Suddenly, Wendel experiences a major loss of thrust with his two burning (gas turbines) under the wing of the Me 262 V1 - Stage 2. Thick dark smoke is billowing out of each. Digital image by Mario Merino.

The hybrid Me 262 V1 - Stage 2 is coming down, however, Messerschmitt AG test pilot Fritz Wendel has it under control only because of the "turning" Junkers Jumo 210G piston engine. Without it the prototype would have certainly crashed. Digital image by Mario Merino.

Messerschmitt Bf 109TL

Port side profile the proposed twin gas turbine powered Bf 109TL fighter. Digital image by Jozef Gatial.

Me 262 V1 Stage 3

Port side digital image of the Me 262 V1 Stage 3 (PC+UB) by Jozef Gatial.

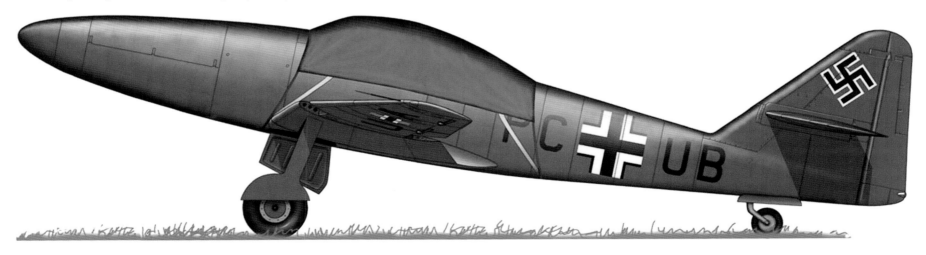

The Me 262 V1 Stage 3. Digital image by Jozef Gatial.

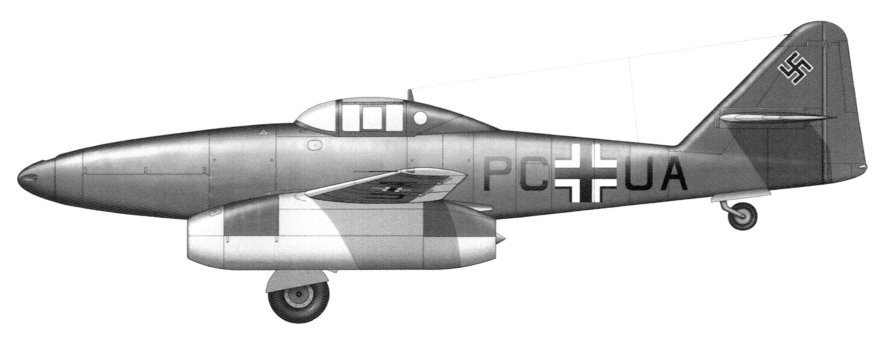

Port side fuselage view of the Me 262 V1 Stage 3 modified for static cockpit pressurization tests since the airframe was no longer airworthy. Notice its tell-tale porthole in the aft cockpit required during cockpit pressurization tests. Digital image by Jozef Gatial.

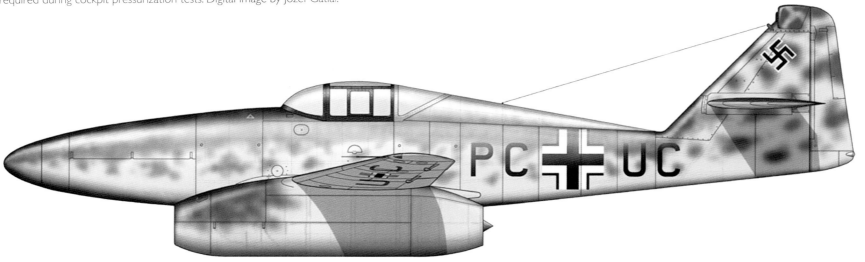

Port side fuselage profile of the Me 262 V1 Stage 3 with the fuselage aft the cockpit smoothly blended in an effort to improve air flow over the tail assembly. The results of this modification is unclear. Digital image by Jozef Gatial.

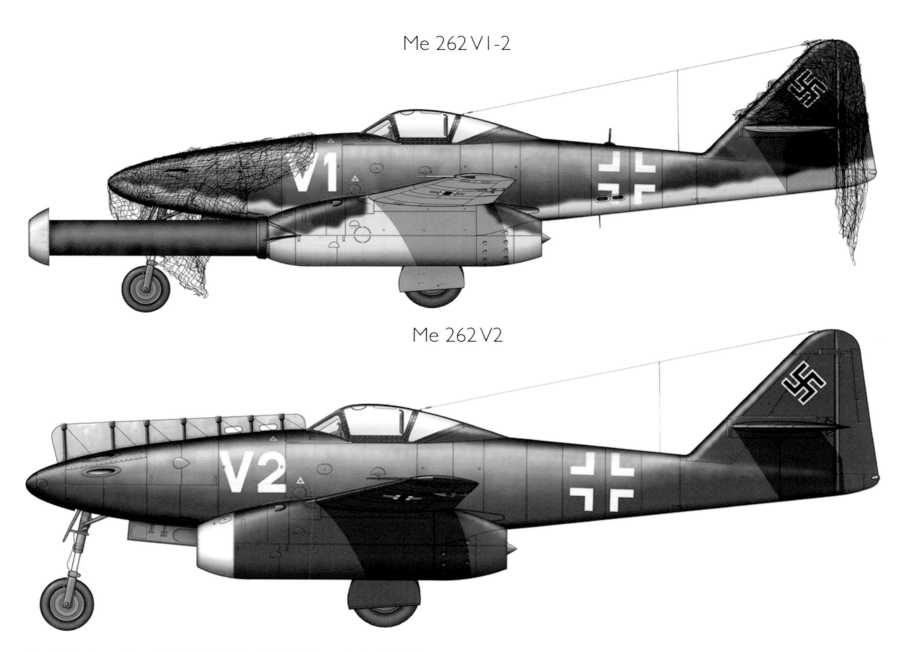

Me 262 V1-2

Me 262 V2

Port side fuselage profiles of the Me 262 V1-2 and V2. Digital images by Jozef Gatial.

Me 262 V2-2 (2nd Version Known As [V056])

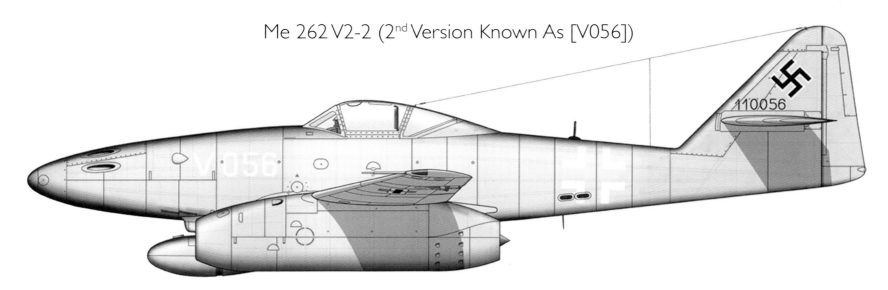

Port side fuselage profile of the Me 262 V2-2. Digital image by Jozef Gatial.

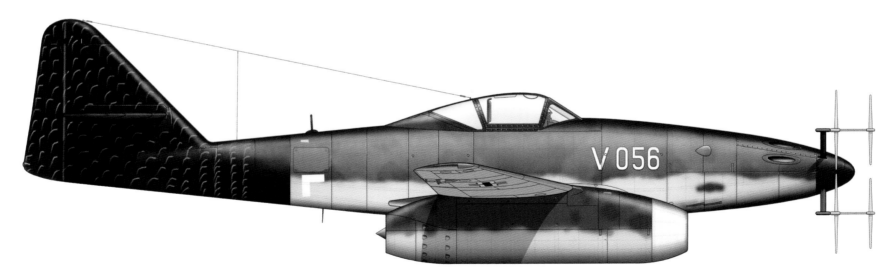

Me 262 V2 2nd version [V056] featuring the installation of a Siemens (Funk Gerät) FuG 218 "Neptun VI" nose mounted radar array. This Me 262 V2-2 FuG 218 version is characterized by a painted black tail assembly, which at one time held cotton tufts to determine if air flow patterns over the tail were changed appreciably due to its speed-robbing FuG 218 radar array. Digital image by Jozef Gatial.

In addition to the two man Me 262 B-1a/U1 night fighter, Messerschmitt AG was also proposing a single seat night fighter version. The designation for the single seat Me 262 night fighter is unclear. Digital image by Jozef Gatial.

Me 262 V2-2 [V056]
with a Siemens FuG 218 Search Radar Array and
Port Wing Lorenz FuG 226 "Neuling" IFF Mast Antenna

The single seat Me 262 V2 2nd version shown with à Siemens FuG 218 search radar and Lorenz FuG 226 "Neuling" IFF dual mast antenna on the port side upper wing. Digital image by Jozef Gatial.

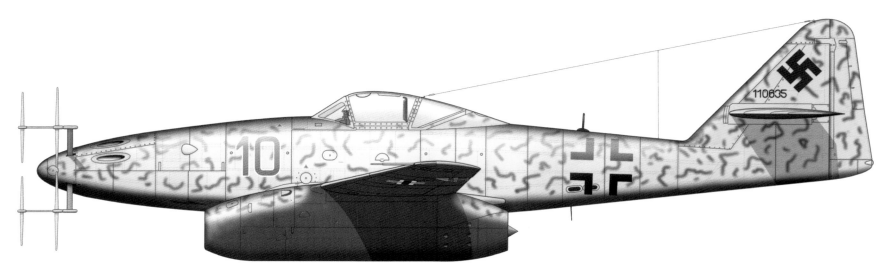

Port side fuselage profile of the proposed single seat night fighter equipped with a Siemens (Funk Gerät) FuG 218 search radar. Digital image by Jozef Gatial.

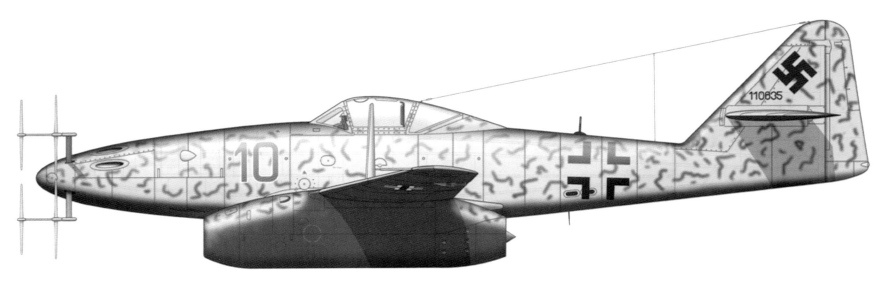

Port side fuselage profile of the proposed single seat Me 262 A-1a night fighter equipped with a Siemens (Funk Gerät) FuG 218 search radar, plus a Lorenz FuG 226 IFF mast antenna on its port wing. Digital image by Jozef Gatial.

Me 262 V2-2 [V056]
with a Siemens (Funk Gerät) FuG 218 Nose Antenna Array and
a Lorenz FuG 226 "Neuling" IFF Port Wing Single Mast Antenna

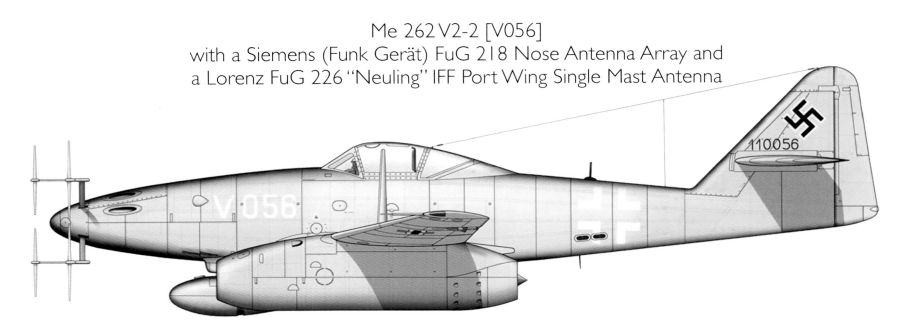

A Me 262 A-1a series production version (based on the Me 262 V2 2nd version) with a nose mounted antenna array for the Lorenz airborne (Funk Gerät) FuG 226 "Neuling" radar plus a single IFF mast antenna on its port side upper wing. Digital image by Jozef Gatial.

Me 262 V2-2 [V056]
with Experimental Dorsal Fin

Me 262 V2 2nd version [V056] with an experimental dorsal fin extending from forward cockpit canopy to the base of the vertical tail plane. Digital image by Jozef Gatial.

Me 262 V2-2 [V056]
Initial Experiments with a Lowered Vertical Stabilizer-Rudder-Trim Tab

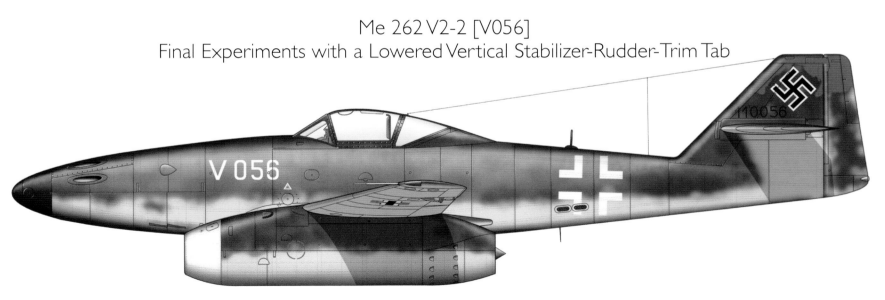

Port side fuselage profile of Me 262 V2-2 [V056] featuring its shortened vertical stabilizer. Digital image by Jozef Gatial.

Me 262 V2-2 [V056]
Final Experiments with a Lowered Vertical Stabilizer-Rudder-Trim Tab

The Me 262 V2 2nd [V056] version with what appears to have been the final modification to the tail plane assembly. Digital image by Jozef Gatial.

Me 262 V3

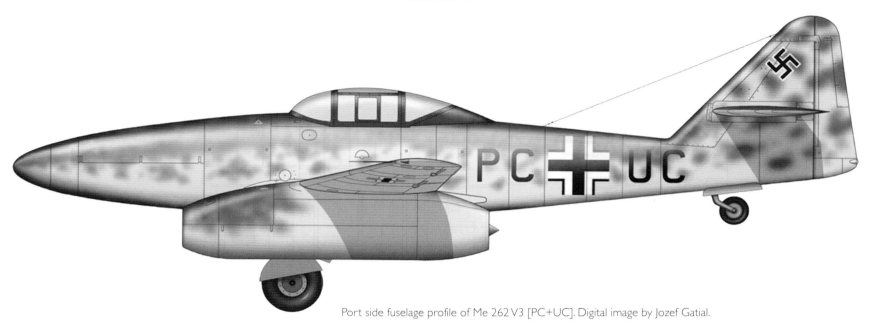

Port side fuselage profile of Me 262 V3 [PC+UC]. Digital image by Jozef Gatial.

Starboard side profile of the Me 262 V3 [PC+UC]. Scale model and photograph by Günter Sengfelder.

Me 262 V4

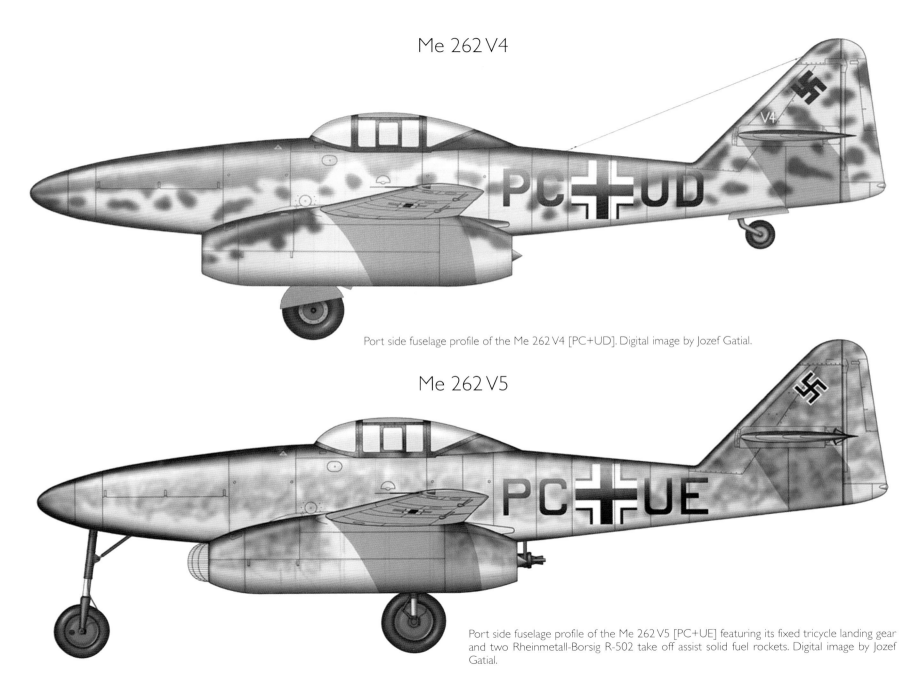

Port side fuselage profile of the Me 262 V4 [PC+UD]. Digital image by Jozef Gatial.

Me 262 V5

Port side fuselage profile of the Me 262 V5 [PC+UE] featuring its fixed tricycle landing gear and two Rheinmetall-Borsig R-502 take off assist solid fuel rockets. Digital image by Jozef Gatial.

Me 262 V5-2 [V167]

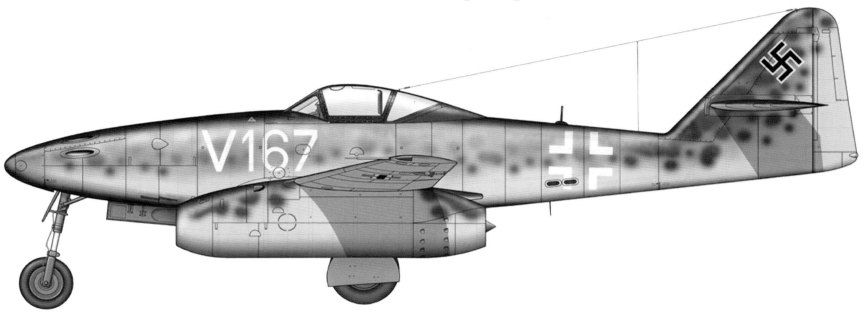

Port side profile image of Me 262 V7-2 [V167]. Digital image by Jozef Gatial.

Port side profile image of Me 262 V7-2 [SQ+WF]. Digital image by Jozef Gatial.

Me 262 V6

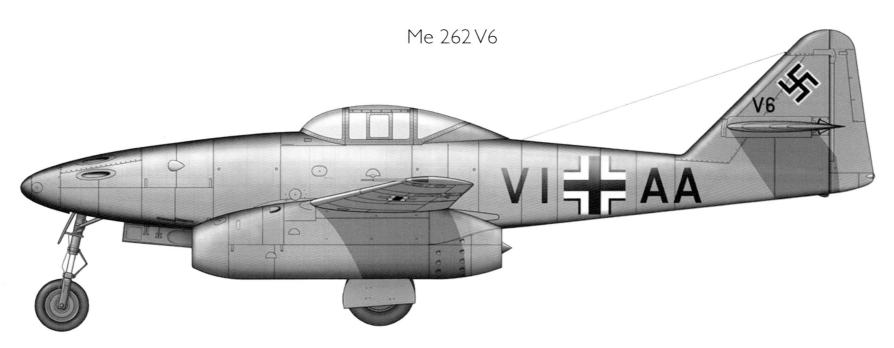

Port side profile of Me 262 V6 [VI+AA] painted in RLM 75 and featuring its old style cockpit canopy. Digital image by Jozef Gatial.

Me 262 V7

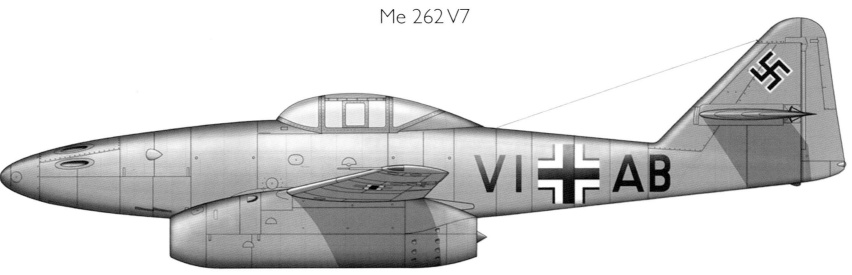

Port side view of Me 262 V7 [VI+AB]. Notice its old style cockpit canopy. Digital image by Jozef Gatial.

Me 262 V7-2

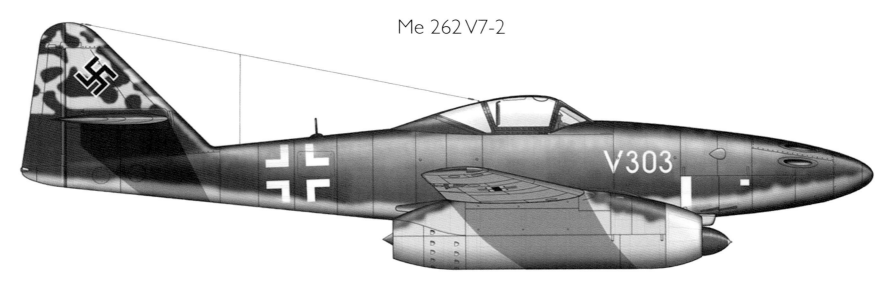

Starboard side profile of the Me 262 V7-2 [V303]. Digital image by Jozef Gatial.

Me 262 V8

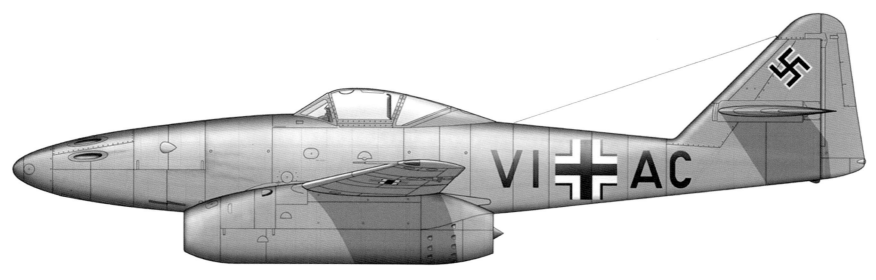

A port side profile of the Me 262 V8 [VI+AC]. Digital image by Jozef Gatial.

Me 262 V8-2 [V484]

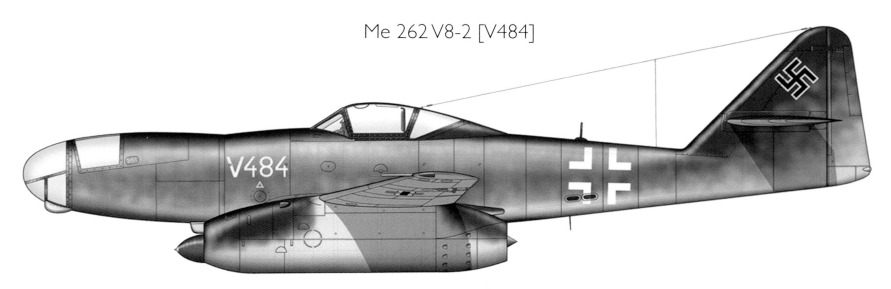

Port side profile of the Me 262 V484. Digital image by Jozef Gatial.

Me 262 V9

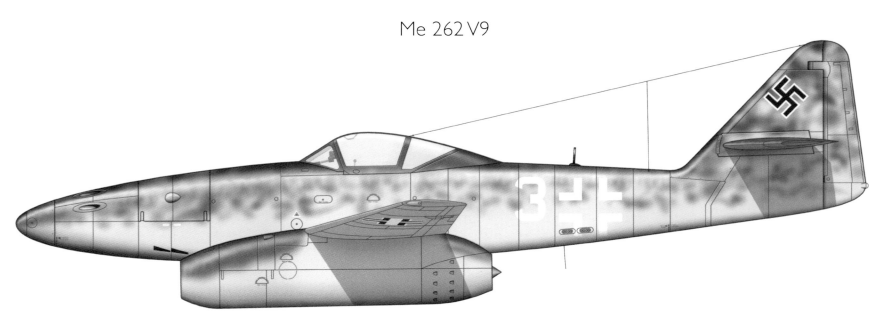

Port side profile of the Me 262 V9. Digital image by Jozef Gatial.

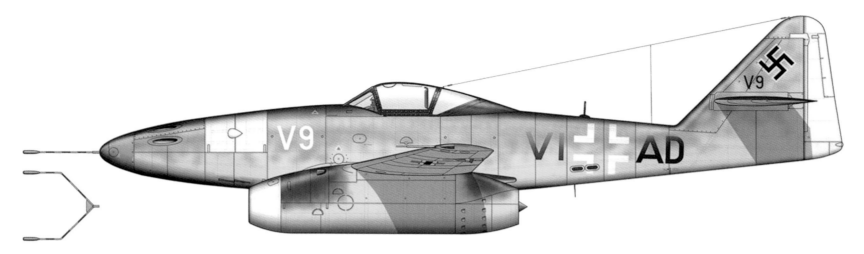

A port side profile of the Me 262 V9 [VI+AD] featuring its experimental acoustical direction finding equipment. Digital image by Jozef Gatial.

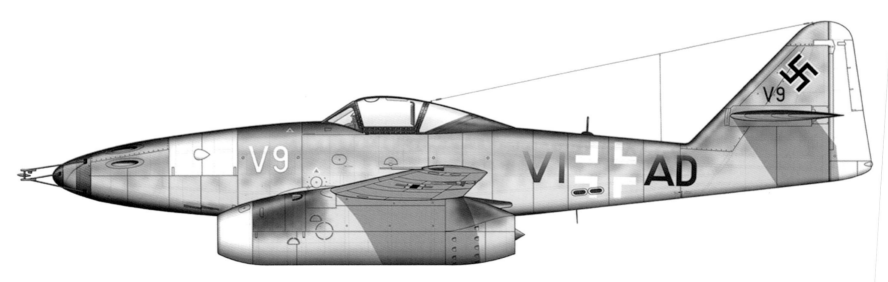

A port side fuselage profile of the Me 262 V9 [VI+AD] featuring its experimental acoustic gathering electric probes. Digital image by Jozef Gatial.

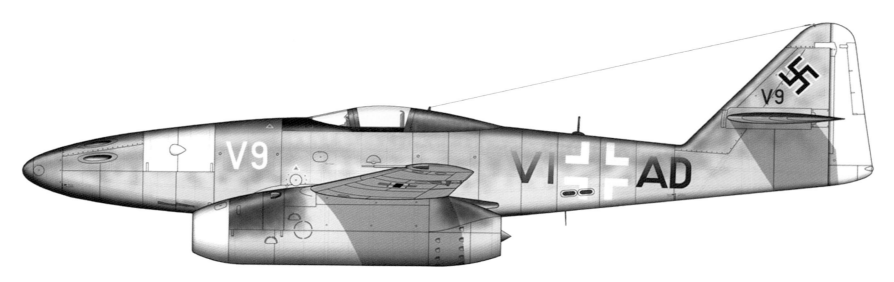

The Me 262 V9 (VI+AD) was later used in high speed test flights. Messerschmitt AG design people redesigned the cockpit wind screen and cockpit canopy to give it a more streamlined profile. These modifications were completed in early October 1944. Messerschmitt AG designers called their new streamlined cabin the "Rennkabine" or racing cabin. Digital image by Jozef Gatial.

Me 262 V10

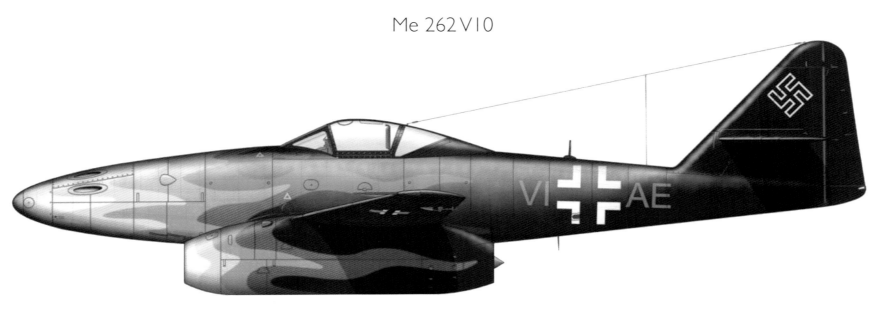

The Me 262 V10 (VI+AE). Digital image by Jozef Gatial.

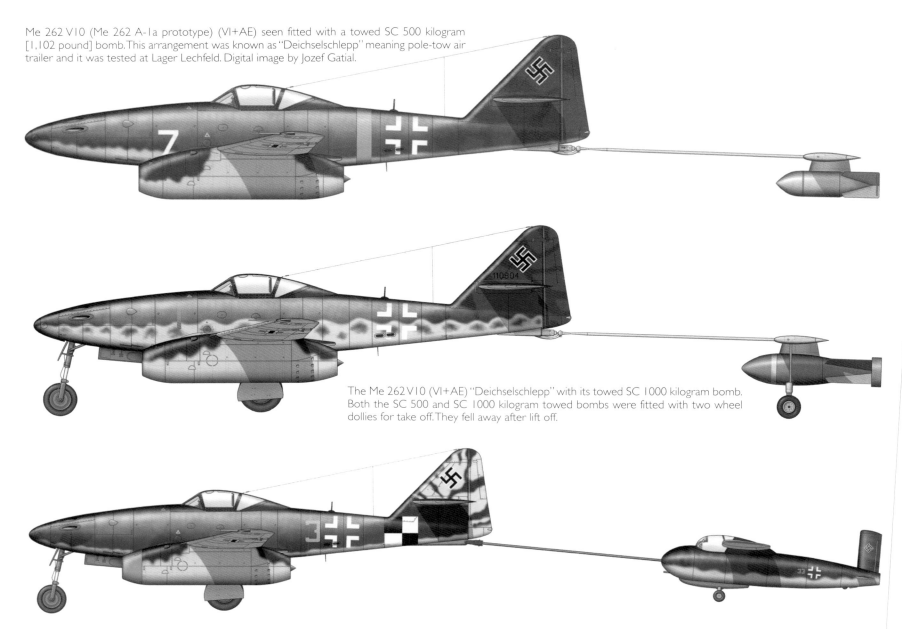

Me 262 V10 (Me 262 A-1a prototype) (VI+AE) seen fitted with a towed SC 500 kilogram [1,102 pound] bomb. This arrangement was known as "Deichselschlepp" meaning pole-tow air trailer and it was tested at Lager Lechfeld. Digital image by Jozef Gatial.

The Me 262 V10 (VI+AE) "Deichselschlepp" with its towed SC 1000 kilogram bomb. Both the SC 500 and SC 1000 kilogram towed bombs were fitted with two wheel dollies for take off. They fell away after lift off.

A proposed idea for the "Deichselschlepp," or pole-towed concept was to tow a Messerschmitt Me P-1103 "Bordjäger" or a pole-towed HWK 509 bi-fuel liquid rocket glider/rammer flying machine behind a Me 262 A-1a. Digital image by Jozef Gatial.

Me 262 V11 [V555]

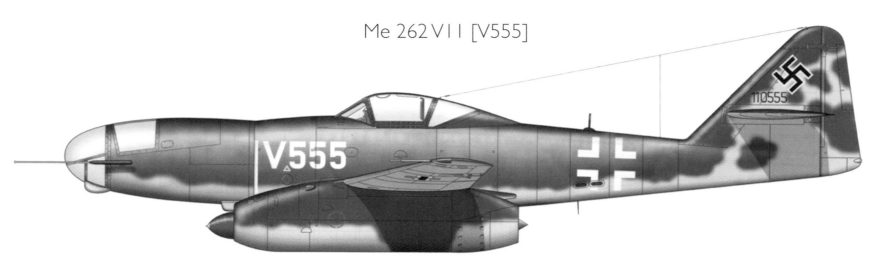

The Me 262 V11 [V555] digital image by Jozef Gatial.

Me 262 V12

Port side digital image of the Me 262 V12 (V074) "Heimatschützer II" or home land defender by Jozef Gatial. The long trailing pipe under the tail assembly is the emergency fuel dump for BMW 718 bi-fuel rocket engine attached to the BMW 003R gas turbine.

Me 262 S-3

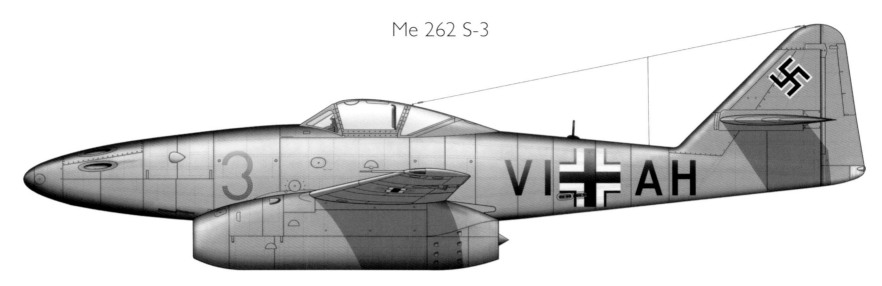

Port side profile of the Me 262 S-3. Digital image by Jozef Gatial.

Me 262 S-4

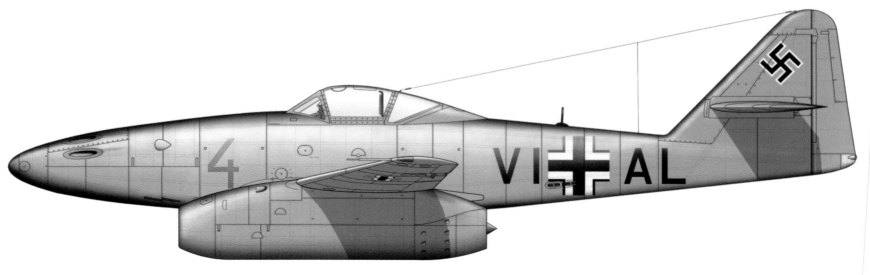

Port side image of the Me 262 S-4. Digital image by Jozef Gatial.

Me 262 S-5

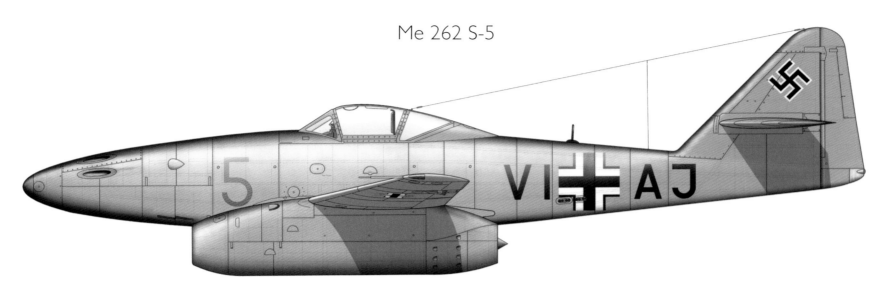

Port side profile image of the Me 262 S-5 VI+AJ. Digital image by Jozef Gatial.

Me 262 S-6

Port side image of the Me 262 S-6. Digital image by Jozef Gatial.

Me 262 S-7

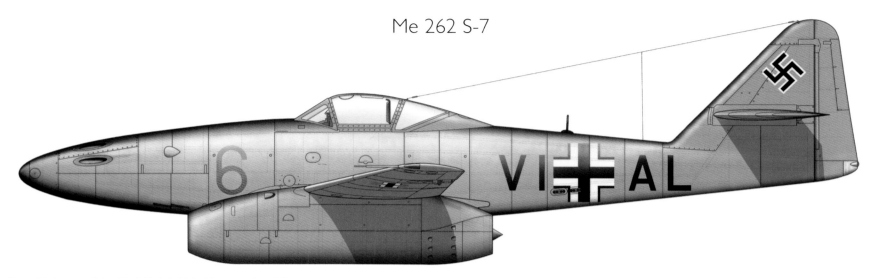

Port side image of the Me 262 S-6. Digital image by Jozef Gatial.

Me 262 S-8

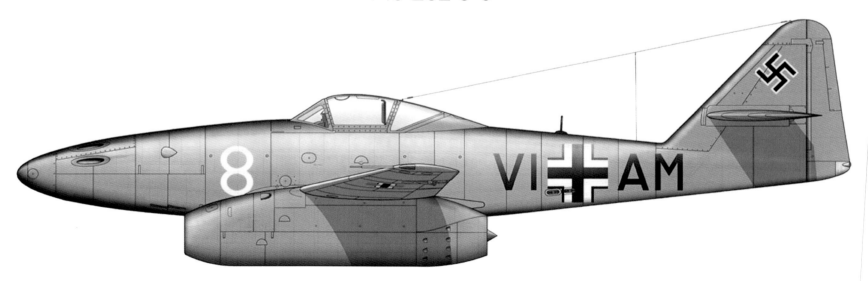

Digital image of the Me 262 S-8 by Jozef Gatial.

Me 262 S-9

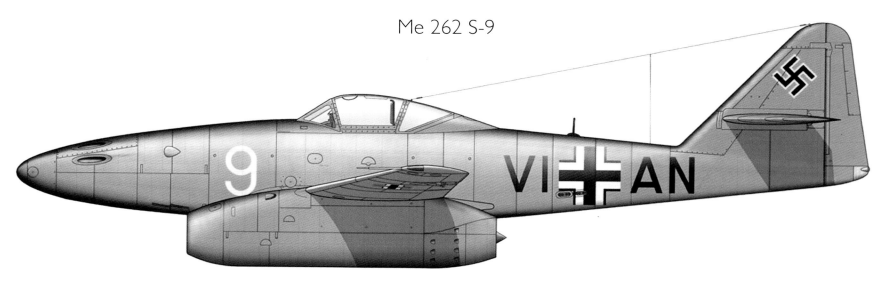

Port side image of the Me 262 S-9. Digital image by Jozef Gatial.

Me 262 S-10

Port side profile of the Me 262 S-10. Digital image by Jozef Gatial.

Me 262 S-11

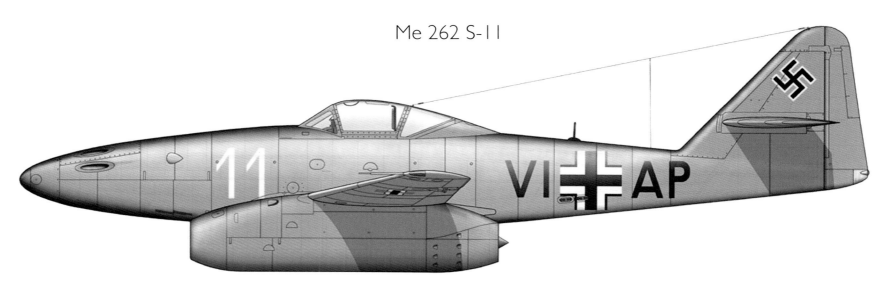

Port side profile of the Me 262 S-11. Digital image by Jozef Gatial.

Me 262 S-12

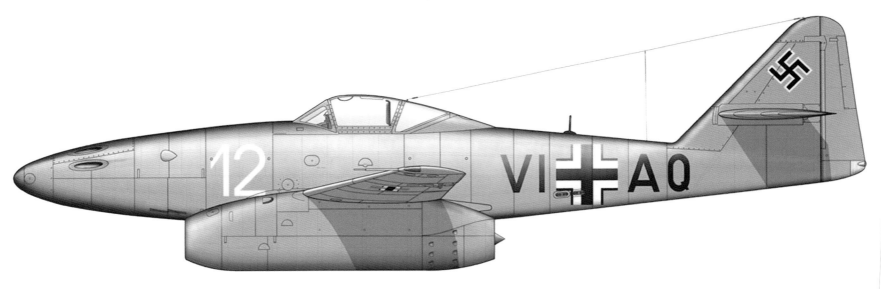

Port side profile image of the Me 262 S-12. Digital by Jozef Gatial.

Me 262 S-13

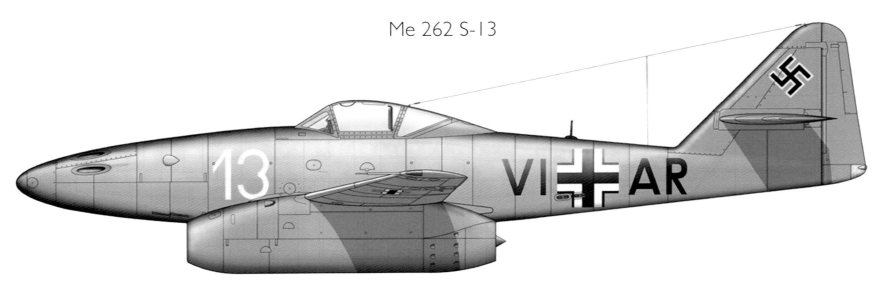

Port side profile of the Me 262 S-13. Digital image by Jozef Gatial.

Me 262 S-14

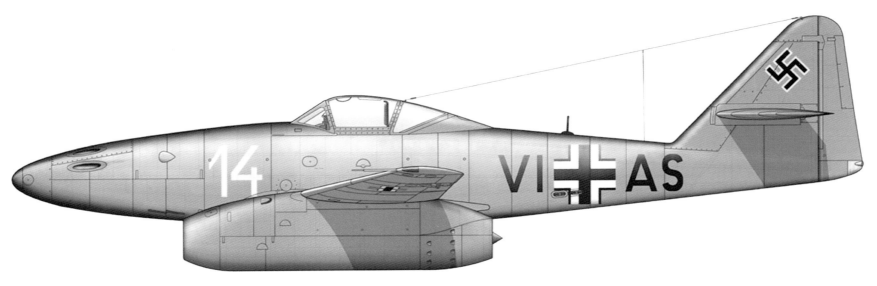

Digital image of the Me 262 S-14 by Jozef Gatial.

Me 262 S-15

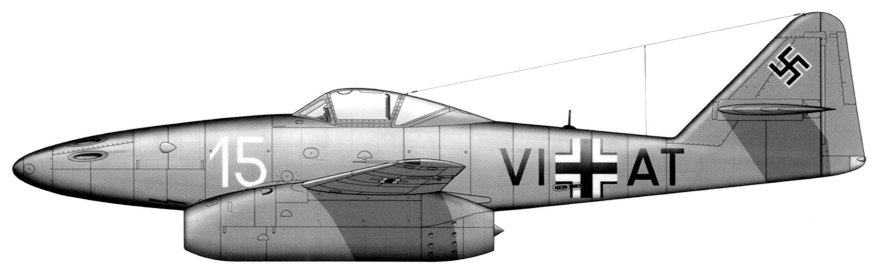

Digital image of the Me 262 S-15 by Jozef Gatial.

Index

X Planes of the Third Reich: Messerschmitt P.1101

David Myhra
Powered by a single HeS 001A turbojet engine, Woldemer Voigt, who had artfully crafted the Me 262, ran out of time before he could make the 1101's design "jell" as he struggled to produce the world's first variable wing sweep, ultra light weight interceptor, and armed with Germany's state-of-the art wing-mounted air-to-air guided missiles. Post-war, Bell Aircraft sought to carry on Voigt's planning and resolved to make the complicated mathematics of light weight, variable wing sweep, and wing-mounted weapons come together in a single aircraft design. The result was the Bell X-5, and it too was disappointing. This photographic history of the Me P.1101 by David Myhra features mostly previously unpublished photos, three-view line drawings, and stunningly realistic photos of a 1101 scale model.
Size: 11" x 8 1/2" • Pages: 64 • Illustrations: over 100 b/w photographs and line drawings
ISBN: 0764309080 • soft cover • $9.95

**Messerschmitt Me 262
Development /Testing/Production**

Willy Radinger & Walter Schick
Accounts of the developments of the fighter, fighter/bomber, reconnaissance, and night fighter versions.
Size: 8 1/2" x 11" • Pages: 112 pages • Illustrations: over 150 b/w, and 30 color photographs, documents
ISBN: 0887405169 • hard cover • $24.95

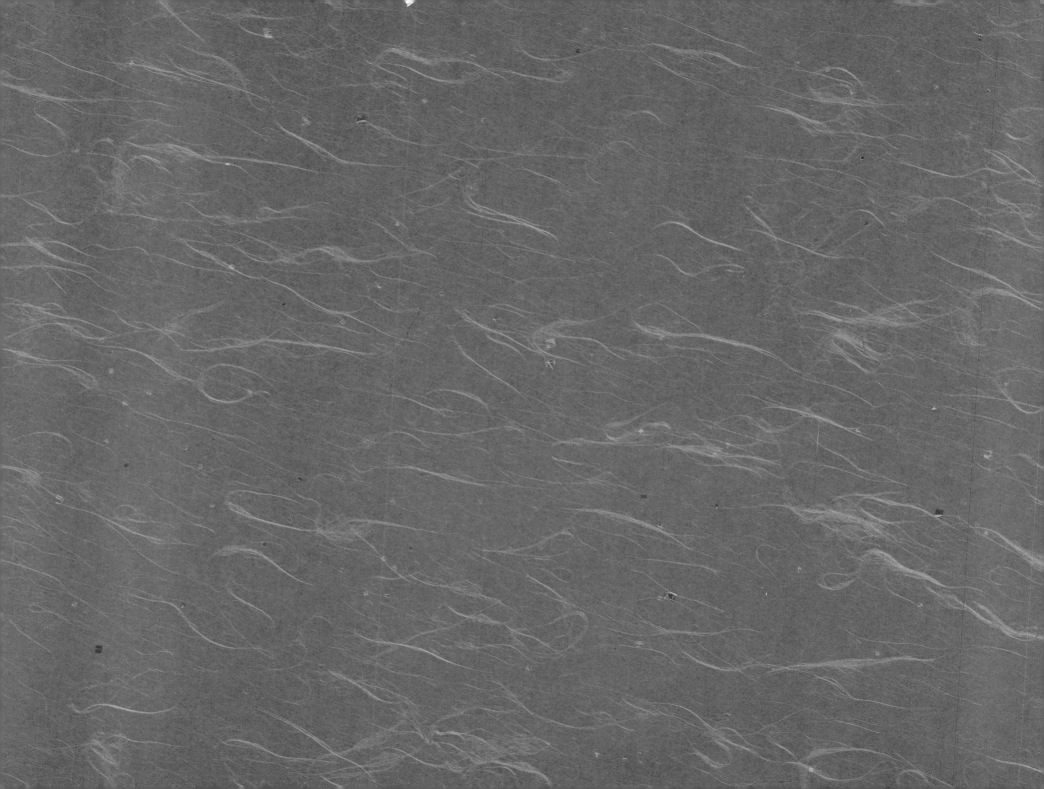